CAMBRIDGE LIBRARY COLLECTION

Books of enduring scholarly value

Darwin

Two hundred years after his birth and 150 years after the publication of 'On the Origin of Species', Charles Darwin and his theories are still the focus of worldwide attention. This series offers not only works by Darwin, but also the writings of his mentors in Cambridge and elsewhere, and a survey of the impassioned scientific, philosophical and theological debates sparked by his 'dangerous idea'.

Physical Geography

Mary Somerville (1780–1872) would have been a remarkable woman in any age, but as an acknowledged leading mathematician and astronomer at a time when the education of most women was extremely restricted, her achievement was extraordinary. Laplace famously told her that 'There have been only three women who have understood me. These are yourself, Mrs Somerville, Caroline Herschel and a Mrs Greig of whom I know nothing.' Mary Somerville was in fact Mrs Greig. After (as she herself said) translating Laplace's work 'from algebra into common language', she wrote On the Connexion of the Physical Sciences (1834), also reissued in this series. Her next book, the two-volume Physical Geography (1848), was a synthesis of geography, geology, botany and zoology, drawing on the most recent discoveries in all these fields to present an overview of current understanding of the natural world and the Earth's place in the universe.

Physical Geography

VOLUME 1

MARY SOMERVILLE

CAMBRIDGE UNIVERSITY PRESS

Cambridge, New York, Melbourne, Madrid, Cape Town, Singapore,
São Paolo, Delhi, Dubai, Tokyo

Published in the United States of America by Cambridge University Press, New York

www.cambridge.org
Information on this title: www.cambridge.org/9781108005203

This edition first published 1848
This digitally printed version 2009

ISBN 978-1-108-00520-3 Paperback

W Holl, sc

Mary Somerville

From a Drawing by James R. Swinton, Esq. 1848.

London, John Murray, Albemarle Street, 1848.

PHYSICAL GEOGRAPHY.

BY

MARY SOMERVILLE,

AUTHOR OF THE "CONNEXION OF THE PHYSICAL SCIENCES,"
"MECHANISM OF THE HEAVENS."

IN TWO VOLUMES.

VOL. I.

WITH A PORTRAIT.

LONDON:
JOHN MURRAY, ALBEMARLE STREET.

1848.

TO

SIR JOHN F. W. HERSCHEL, BART., K.H.,

&c. &c.

DEAR SIR JOHN,

I AVAIL myself with pleasure of your permission to dedicate my book to you, as it gives me an opportunity of expressing my admiration of your talents, and my sincere estimation of your friendship.

I remain, with great regard,

Yours truly,

MARY SOMERVILLE.

London, 29th February, 1848.

CONTENTS OF VOL. I.

CHAPTER I.

CHAPTER II.

CHAPTER III.

CHAPTER IV.

CHAPTER V.

CHAPTER VI.

CHAPTER VII.

CHAPTER VIII.

CHAPTER IX.

CHAPTER X.

CHAPTER XI.

CHAPTER XII.

CHAPTER XIII.

CHAPTER XIV.

CHAPTER XV.

CHAPTER XVI.

CHAPTER XVII.

PHYSICAL GEOGRAPHY.

CHAPTER I.

GEOLOGY.

THE change produced in the civilized world within a few years, by the application of the powers of nature to locomotion, is so astonishing, that it leads to a consideration of the influence of man on the material world, his relation with regard to animate and inanimate beings, and the causes which have had the greatest effect on the physical, moral, and intellectual condition of the human race.

The former state of our terrestrial habitation, the successive convulsions which have ultimately led to its present geographical arrangement, and to the actual distribution of land and water, so powerfully influential on the destinies of mankind, are circumstances of primary importance.

The position of the earth with regard to the sun, its connexion with the bodies of the solar system, together with its size and form, have been noticed by the author elsewhere. It was there shown that our globe forms but an atom in the immensity of space, utterly invisible from the nearest fixed star, and scarcely a telescopic object to the remote planets of our own system. The increase of temperature with the depth below the surface of the earth, and the tremendous desolation hurled over wide regions by numerous fire-breathing mountains, show that man is removed but a few miles from immense lakes or seas of liquid fire. The very shell on which he stands is unstable under his feet, not only from those temporary convulsions that seem to shake the globe to its centre, but from a slow almost imperceptible elevation in some places, and an equally gentle subsidence in others, as if the internal molten matter were subject to secular tides, now heaving and now ebbing, or that the subjacent rocks were in one place expanded and in another contracted by changes of temperature.

The earthquake and the torrent, the august and terrible ministers of Almighty power, have torn the solid earth and opened the seals of the most ancient records of creation, written in indelible characters on "the perpetual hills, and the everlasting mountains." There we read of the changes that have brought the rude mass to its present fair state, and of the myriads of beings that have appeared on this mortal stage, have fulfilled their destinies, and have been swept from existence to make way for new races which, in their turn, have vanished from the scene till the creation of man completed the glorious work. Who shall define the periods of those mornings and evenings when God saw that his work was good? and who shall declare the time allotted to the human race, when the generations of the most insignificant insect existed for unnumbered ages? Yet man is also to vanish in the ever-changing course of events. The earth is to be burnt up, and the elements are to melt with fervent heat—to be again reduced to chaos—possibly to be renovated and adorned for other races of beings. These stupendous changes may be but cycles in those great laws of the universe, where all is variable but the laws themselves and He who has ordained them.

The earth consists of a great variety of substances, some of which occur in amorphous masses, others are disposed in regular layers or strata, either horizontal or inclined at all angles to the horizon. By mining, man has penetrated only a very little way, but by reasoning from the dip or inclination of the strata at or near the surface, and from other circumstances, he has obtained a pretty accurate idea of the structure of our globe to the depth of about ten miles. All the substances of which we have any information are divided into four classes, distinguished by the manner in which they have been formed, namely—Plutonic and Volcanic rocks, both of igneous origin, though produced under different circumstances; Aqueous or Stratified rocks, entirely due to the action of water, as the name implies; and Metamorphic rocks, deposited also by water, according to the opinion of many eminent geologists, and consequently stratified, but subsequently altered and crystallized by heat. The Aqueous and Volcanic rocks are formed at the surface of the earth, the Plutonic and Metamorphic at great depths, but all of them have originated simultaneously during every geological period, and are now in a state of slow and constant progress. The antagonist principles of fire and water have ever been and still are the cause of the perpetual vicissitudes to which the crust of the earth is liable.

It has been ascertained by observation that the Plutonic rocks, consisting of the granites and some of the porphyries, were formed in the deep and fiery caverns of the earth, of melted matter, which crystallized as it slowly cooled under enormous pressure, and was then heaved in unstratified masses by the elastic force of the internal heat even to the tops of the highest mountains, or forced in a semifluid state into fissures of the superincumbent strata, sometimes into the cracks of previously formed granite; for that rock, which constitutes the base of so large a portion of the earth's crust, has not been all formed at once; some portions had been solid while others were yet in a liquid state. This class of rocks is completely destitute of fossil remains.

Although granite and the volcanic rocks are both due to the action of fire, their nature and position are very different: granite, fused in the interior of the earth, has been cooled and consolidated before coming to the surface; besides, it generally consists of few ingredients, so that it has nearly the same character in all countries. But as the volcanic fire rises to the very surface of the earth, fusing whatever it meets with, volcanic rocks take various forms, not only from the different kinds of strata which are melted, but from the different conditions under which the liquid matter has been cooled, though most frequently on the surface—a circumstance that seems to have had the greatest effect on its appearance and structure. Sometimes it approaches so nearly to granite that it is difficult to perceive a distinction; at other times it becomes glass: in short, all those massive, unstratified, and occasionally columnar rocks, as basalt, greenstone, porphyry, and serpentine, are due to volcanic fires, and are devoid of fossil remains.

There seems scarcely to have been any age of the world in which volcanic eruptions have not taken place in some part of the globe. Lava has pierced through every description of rocks, spread over the surface of those existing at the time, filled their crevices, and flowed between their strata. Ever changing its place of action, it has burst out at the bottom of the sea as well as on dry land. Enormous quantities of scoriæ and ashes have been ejected from numberless craters, and have formed extensive deposits in the sea, in lakes, and on the land, in which are imbedded the remains of the animals and vegetables of the epoch. Some of these deposits have become hard rock, others remain in a crumbling state; and as they alternate with the aqueous strata of almost every period, they contain the fossils of all the geological epochs, chiefly fresh and salt water testaceæ.

According to a theory now generally adopted, which originated with Mr. Lyell, whose works are models of philosophical investigation, the metamorphic rocks, which consist of gneiss, mica-schist, clay-slate, statuary marble, &c., were formed of the sediment of water in regular layers, differing in kind and colour, but, having been deposited near the places where plutonic rocks were generated, they have been changed by the heat transmitted from the fused matter, and in cooling under heavy pressure and at great depths they have become as highly crystallized as the granite itself, without losing their stratified form. An earthy stratum has sometimes been changed into a highly crystallized rock to the distance of a quarter of a mile from the point of contact by transmitted heat, and there are instances of dark-coloured limestone full of fossil shells, that has been changed into statuary marble from that cause. Such alterations may frequently be seen to a small extent in rocks adjacent to a stream of lava. There is not a trace of organic remains in the metamorphic rocks; their strata are sometimes horizontal, but they are usually tilted at all angles to the horizon, and form some of the highest mountains and most extensive table-lands on the face of the globe. Although there is the greatest similarity in the plutonic rocks in all parts of the world, they are by no means identical; they differ in colour, and even in ingredients, though these are few.

Aqueous rocks are all stratified, being the sedimentary deposits of water. They originate in the wear of the land by rain, streams, or the ocean. The débris carried by running water is deposited at the bottom of the seas and lakes, where it is consolidated, and then raised up by subterraneous forces, again to undergo the same process after a lapse of time. By the washing away of the land the lower rocks are laid bare, and, as the materials are deposited in different places according to their weight, the strata are exceedingly varied, but consist chiefly of arenaceous or sandstone rocks, argillaceous or clayey rocks, and of calcareous rocks composed of sand, clay, and carbonate of lime. They constitute three great classes, which, in an ascending order, are the primary and secondary fossiliferous strata, and the Tertiary formations.

The primary fossiliferous strata, the most ancient of all the sedimentary rocks, consisting of limestone, sandstones, and shales, are entirely of marine origin, having been formed far from land at the bottom of a very deep ocean; consequently they contain the exuviæ of marine animals only, and after the lapse of unnumbered ages the ripple marks of the waves are still distinctly visible on some of their strata. This series of rocks is subdivided into the Cambrian and the upper and lower Silurian systems, on account of differences in their fossil remains.

The Cambrian rocks, sometimes many thousand yards thick, are for the most part destitute of organic remains, but the Silurian rocks abound in them more and more as the strata lie higher in the series. In the lower Silurian group are the remains of shell-fish, almost all of extinct genera, and the few that have any affinity to those alive are of extinct species;

Crinoidea, or stone-lilies, which had been fixed to the rocks like tulips on their stems, are coëval with the earliest inhabitants of the deep; and the trilobite, a jointed creature of the crab kind, with prominent eyes, are almost exclusively confined to the Silurian strata, but the last traces of them are found in the coal-measures above. In the upper Silurian group are abundance of marine shells of almost every order, together with Crinoidea, vast quantities of corals, and some sea-weeds: several fossil sauroid fish, of extinct genera, but high organization, have been found in the highest beds—the only vertebrated animal that has yet been discovered among the countless profusion of the lower orders of creatures that are entombed in the primary fossiliferous strata. The remains of one or more land-plants, in a very imperfect state, have been found in the Silurian rocks of North America, which shows that there had been land with vegetation at that early period. The type of these plants, as well as the size of the shells and the quantity of the coral, indicate that a uniformly warm temperature had then prevailed over the globe. During the Silurian period an ocean covered the northern hemisphere, islands and lands of moderate size had just begun to rise, and earthquakes with volcanic eruptions, from insular and submarine volcanos, were frequent towards its close.

The secondary fossiliferous strata, which comprise a great geological period, and constitute the principal part of the high land in Europe, were deposited at the bottom of an ocean, like the primary, from the débris of all the others carried down by water, and still bear innumerable tokens of their marine origin, although they have for ages formed part of the dry land. Calcareous rocks are more abundant in these strata than in the crystalline, probably because the carbonic acid was then, as it still is, driven off from the lower strata by the internal heat, and came to the surface as gas or in calcareous springs, which either rose in the sea, and furnished materials for shell-fish and coral insects to build their habitations and form coral reefs, or deposited their calcareous matter on the land in the form of rocks.

The Devonian or old red sandstone group, in many places ten thousand feet thick, consisting of strata of dark red and other sandstones, marls, coralline limestones, conglomerates, &c., is the lowest of the secondary fossiliferous strata, and forms a link between them and the Silurian rocks by an analogy in their fossil remains. It has fossils peculiarly its own, but it has also some shells and corals common to the strata both above and below it. There are various families of extinct sauroid fish in this group, some of which were gigantic, others had strong bony shields on their heads, and one genus, covered with enamelled scales, had appendages like wings. The shark approaches nearer to some of these ancient fish than any other now living.

During the long period of perfect tranquillity that prevailed after the Devonian group was deposited, a very warm, moist, and extremely equable climate, which extended all over the globe, had clothed the islands and lands in the ocean then covering the northern hemisphere with exuberant tropical forests and jungles. Subsequent inroads of fresh water or of the sea, or rather partial sinkings of the land, had submerged these forests and jungles, which, being mixed with layers of sand and mud, had in time been consolidated into one mass, and were then either left dry by the retreat of the waters, or gently raised above their surface.

These constitute the remarkable group of the carboniferous strata, which consists of numberless layers of various substances filled with a prodigious quantity of the remains of fossil land-plants, intermixed with beds of coal, which is entirely composed of vegetable matter. In some cases the plants appear to have been carried down by floods and deposited in estuaries, but in most instances the beauty, delicacy, and sharpness of the impressions show that they had grown on the spot where the coal was formed. More than three hundred fossil plants have been collected from the shale where they abound, frequently with their seeds and fruits, so that enough remains to show the peculiar nature of this flora, whose distinguishing feature was the preponderance of ferns: among these there were tree-ferns which must have been forty or fifty feet high. There were also plants resembling the horse-tail tribe, of gigantic size; others like the tropical club mosses: an aquatic plant of an extinct family was very abundant, beside many others to which we have nothing analogous. Forest-trees of great magnitude, of the pine and fir tribes, flourished at that period. The remains of an extinct araucaria, one of the largest of the pine family, have been found in the British coal-fields; the existing species now grow in very warm countries: a few rare instances occur of grasses, palms, and liliaceous plants. The botanical districts were very extensive when the coal-plants were growing, for the species are nearly identical throughout the coal-fields of Europe and America. From the extent of the ocean, the insular structure of the land, the profusion of ferns and fir-trees, and the warm, moist, and equable climate, the northern hemisphere during the formation of the coal strata is thought to have borne a strong resemblance to the South Pacific, with its fern and fir clothed lands of New Zealand, Kerguelen islands, and others.

The animal remains of this period are in the mountain limestone, a rock occasionally nine hundred feet thick, which, in some instances, lies beneath the coal-measures, and sometimes alternates with the shale and sandstone. They consist of crinoidea and marine testaceæ, among which the size of the chambered shells, as well as that of the corals, shows that the ocean was very warm at that time, even in the high northern latitudes.

The coal strata have been very much broken and deranged in many places by earthquakes, which frequently occurred during the secondary fossiliferous period, and from time to time raised islands and land from the deep. However, these and all other changes that have taken place on the earth have been gradual and partial, whether brought about by fire or water. The older rocks are more shattered by earthquakes than the newer, because the movement came from below; but these convulsions have never extended all over the earth at the same time—they have always been local: for example, the Silurian strata have been dislocated and tossed in Britain, while a vast area in the south of Sweden and Russia still retains a horizontal position. There is no proof that any mountain-chain has ever been raised at once; on the contrary, the elevation has always been produced by a long-continued and reiterated succession of internal convulsions, with intervals of repose. In many instances the land has risen up or sunk down by an imperceptible equable motion continued for ages, while in other places the surface of the earth has remained stationary for long geological periods.

The magnesian limestone, or permian formation, comes immediately above the coal-measures, and consists of breccias or conglomerates, gypsum, sandstone, marl, &c.; but its distinguishing feature is a yellow limestone rock, containing carbonate of magnesia, which often takes a granular texture, and is then known as dolomite. The permian formation has a fossil flora and fauna peculiar to itself, mingled with those of the coal strata. Here the remnant of an earlier creation gradually tends to its final extinction, and a new one begins to appear. The flora is, in many instances, specifically the same with that in the coal strata below. Certain fish are also common to the two, which never appear again. They belong to a race universal in the early geological periods, and bear a strong resemblance to saurian reptiles.

A small number of existing genera only, such as the shark and sturgeon, make some approach to the structure of these ancient inhabitants of the waters. The new creation is marked by the introduction of two species of saurian reptiles: the fossil remains of one have been found in the magnesian limestone in England, and those of the other in a corresponding formation in Germany. They are the earliest members of a family which was to have dominion in the land and water for ages.

A series of red marls, rock-salt, and sandstones, which have arisen from the disintegration of metamorphic slates and porphyritic trap containing oxide of iron, and known as the trias or new red sandstone system, lies above the magnesian limestone. In England this formation is particularly rich in rock-salt, which, with layers of gypsum and marl, is sometimes six hundred feet thick; but in this country the muschelkalk is wanting, which in Germany is so remarkable for the quantity of organic remains. At this time creatures like frogs of enormous dimensions had been frequent, as they have left their footsteps on what must then have been a soft shore. Forty-seven genera of fossil remains have been found in the trias in Germany, consisting of shells, cartila ginous fish, encrinites, &c., all distinct in species, and many distinct in genera, from the organic fossils of the magnesian limestone below, and also from those entombed in the strata above.

During a long period of tranquillity the oolite or jurassic group was next deposited in a sea of variable depth, and consisted of sands, sandstones, marls, clays, and limestone. At this time there was a complete change in the aqueous deposits all over Europe. The red iron-stained arenaceous rocks, the black coal, and dark strata were succeeded by light blue clays, pale yellow limestones, and, lastly, white chalk. The water that deposited the strata must have been highly charged with carbonate of lime, since few of the formations of that period are without calcareous matter, and calcareous rocks were formed to a prodigious extent throughout Europe; the Pyrenees, Alps, Apennines, and Balkan abound in them, and the Jura mountains, which have given their name to the series, are formed of them. The European ocean then teemed with animal life; whole beds consist almost entirely of marine shells and corals. Belemnites and ammonites, from an inch in diameter to the size of a cart-wheel, are entombed by myriads in the strata; whole forests of that beautiful zoophite, the stone-lily, flourished on the surface of the oolite, then under the waters; and the encrinite, one of the same genus, is embedded in millions in the enchoreal shell marble, which occupies such extensive tracts in Europe. Fossil fish are numerous in these strata, but different from those of the coal series, the permian formation, and trias. Not one genus of the fish of this period are now in existence. The newly-raised islands and lands were clothed with vegetation like that of the large islands of the intertropical Archipelagos of the present day, which, though less rich than during the carboniferous period, still indicates a very moist and warm climate. Ferns were less abundant, and they were associated with various genera and species of the cycadeæ, which had grown on the southern coast of England, and in other parts of northern Europe, congeners of the present cycas and zamia of the tropics. These plants had been very numerous, and the pandanæ, or screw-pine, the first tenant of the new lands in ancient and modern times, is a family found in a fossil state in the inferior oolite of England, which was but just rising from the deep at that time. The species now flourishing grows only on the coasts of such coral islands in the Pacific as have recently emerged from the waves. In the upper strata of this group, however, the con fervæ and monocotyledonous plants become more rare—an indication of a change of climate.

The new lands that were scattered in the ocean of the oolitic period were drained by rivers, and inhabited by huge crocodiles and saurian reptiles of gigantic size, mostly of extinct genera. The crocodiles came nearest to modern reptiles, but the others, though bearing a remote similitude in general structure to living forms, were quite anomalous, combining in one the structure of various distinct creatures, and so monstrous that they must have been more like the visions of a troubled dream than things of real existence; yet in organization a few of them came nearer to the type of living mammalia than any existing reptiles do. Some of these saurians had lived in the water, others were amphibious, and the various species of one genus even had wings like a bat, and fed on insects. There were both herbivorous and predaceous saurians, and from their size and strength they must have been formidable enemies. Besides, the numbers deposited are so great that they must have swarmed for ages in the estuaries and shallow seas of the period, especially in the lias, a marine stratum of clay the lowest of the oolite series. They gradually declined towards the end of the secondary fossiliferous epoch, but as a class they lived in all subsequent eras, and still exist in tropical countries, although the species are very different from their ancient congeners. Tortoises of various kinds were contemporary with the saurians, also a family that still exists. In the stonefield slate, a stratum of the lower oolitic group, there are the remains of insects ; and the bones of two small quadrupeds have been found there belonging to the marsupial tribe, such as the opossum ; a very remarkable circumstance, because that family of animals at the present time is confined to New Holland, South America, and as far north as Pennsylvania at least. The great changes in animal life during this period were indications of the successive alterations that had taken place on the earth's surface.

The cretaceous strata follow the oolite in ascending order, consisting of clay, green and iron sands, blue limestone, and chalk, probably formed of the decay of coral and shells, which predominates so much in England and other parts of Europe, that it has given the name and its peculiar feature to the whole group.

It is, however, by no means universal; the chalk is wanting in many parts of the world where the other strata of this series prevail, and then their connexion with the group can only be ascertained by the identity of their fossil remains. With the exception of some beds of coal among the oolitic series, the Wealden clay, the lowest of the cretaceous group in England, is a fresh-water formation, and the tropical character of its flora shows that the climate was still very warm. Plants allied to the zamias and cycades of our tropical regions, many ferns and pines of the genus araucaria, characterized its vegetation, and the upright stems of a fossil forest at Portland show that it had been covered with trees. It was inhabited by tortoises approaching to families now living in warm countries, and saurian reptiles of five different genera swarmed in the lakes and estuaries. This clay contains fresh-water shells, fish of the carp kind, and the bones of wading birds. The Wealden clay is one of the various instances of the subsidence of land, of which there were others during this period.

The cretaceous strata above our Wealden clay are full of marine exuviæ. There are vast tracts of sand in northern Europe, and many very extensive tracts of chalk, but in the southern part of the continent the cretaceous rocks assume a different character. There and elsewhere extensive limestone rocks, filled with very peculiar shells, show that when the cretaceous strata were forming an ocean extended from the Atlantic into Asia, which covered the south of France, all southern Europe, part of

Syria, the isles of the Ægean Sea, and the coasts of Thrace and the Troad. The remains of turtles have been found in the cretaceous group, quantities of coral, and abundance of shells of extinct species; some of the older kinds still existed, new ones were introduced, and some of the most minute species of microscopic shells, which constitute a large portion of the chalk, are supposed to be the same with creatures now alive, the first instance of identity of species in the ancient and modern creation. An approximation to recent times is to be observed also in the arrangement of organized nature, since at this early period, and indeed even in the silurian and oolitic epochs, the marine fauna was divided, as now, into distinct geographical provinces. The great saurians were on the decline, and many of them were found no more, but a gigantic creature, intermediate between the living monitor and iguana, lived at this period.

An immense geological cycle elapsed between the termination of the secondary fossiliferous strata and the beginning of the tertiary. With the latter a new order of things commenced approaching more closely to the actual state of the globe. During the tertiary formation the same causes under new circumstances produced an infinite variety in the order and kind of the strata, accompanied by a corresponding change in animal and vegetable life. The old creation, which had nothing in common with the existing order of things, had passed away and given place to one more nearly approaching to that which now prevails. Among the myriads of beings that inhabited the earth and the ocean during the secondary fossiliferous epoch scarcely one species is to be found in the tertiary. Two planets could hardly differ more in their natural productions. This break in the law of continuity is the more remarkable, as hitherto some of the newly created animals were always introduced before the older were extinguished. The circumstances and climate suited to the one became more and more unfit for the other, which consequently perished gradually while their successors increased. It is possible that as observations become more extended this hiatus may be filled up.

The series of rocks from the granite to the end of the secondary fossiliferous strata, taken as a whole, constitute the solid crust of the globe, and in that sense are universally diffused over the earth's surface. The tertiary strata occupy the hollows formed in this crust, whether by subterraneous movements, by lakes, or denudation by water, as in the estuaries of rivers, and consequently occur in irregular tracts, often, however, of prodigious thickness and extent. Indeed they seem to have been as widely developed as any other formation, though time has been wanting to bring them into view.

The innumerable basins and hollows with which the continents and larger islands had been indented for ages after the termination of the secondary fossiliferous series, had sometimes been fresh-water lakes, and at other times were inundated by the sea;

consequently the deposits which took place during these changes alternately contain the spoils of terrestrial and marine animals. The frequent intrusion of volcanic strata among the tertiary formations shows that, in Europe, the earth had been in a very disturbed state, and that these repeated vicissitudes had been occasioned by elevations and depressions of the soil, as well as by the action of water.

There are three distinct groups in these strata: the lowest tertiary or Eiocene group, so called by Mr. Lyell, because, among the myriads of fossil shell-fish it contains, very few are identical with those now living; the Meiocene, or middle group, has a greater number of the exuviæ of existing species of shells; and the Pleiocene, or upper tertiary group, still more. Though frequently heaved up to great elevations on the flanks of the mountain-chains, as, for example, on the Alps and Apennines, by far the greater part of the tertiary strata maintain their original horizontal position in the very places where they were formed. Immense insulated deposits of this kind are to be met with all over the world; Europe abounds with them, London and Paris stand on such basins, and they cover immense tracts both in North and South America.

The monstrous reptiles had mostly disappeared, and the mammalia now took possession of the earth, of forms scarcely less anomalous than their predecessors, though approaching more nearly to those alive.

Numerous species of extinct animals that lived during the earliest or Eiocene period have been found in various parts of the world, especially in the Paris basin, of the order of Pachydermata, to the greater number of which we have nothing analogous; they were mostly amphibious and herbivorous quadrupeds, which had frequented the borders of the rivers and lakes that covered the greater part of Europe at that time. This is the more extraordinary, as existing animals of that order, namely, a daman and three tapirs, are confined to the torrid zone. These creatures were widely diffused, and some of them were associated with genera still existing, though of totally different species; such as animals allied to the racoon and dormouse, the ox, bear, deer, the fox, the dog, and others. Although these quadrupeds differ so widely from those of the present day, the same proportion existed then as now between the carnivorous and herbaceous genera. The spoils of marine mammalia of this period have also been found, sometimes at great elevations above the sea, all of extinct species, and some of these cetacea were of huge size. This marvellous change of the creative power was not confined to the earth and the ocean; the air also was now occupied by many extinct races of birds allied to the owl, buzzard, quail, curlew, &c. The climate must still have been warmer than at present from the remains of land and sea plants found in high latitudes. Even in England bones of the opossum, monkey, and boa have been discovered, all animals of warm countries, besides fossil sword and saw fish, both of genera foreign to the British seas.

During the Meiocene period new amphibious quadrupeds were associated with the old, of which the deinotherium is the most characteristic, and much the largest of the mammalia yet found, far surpassing the largest elephant in size, of a singular form, and unknown nature.

The palæotherium was also of this period, and also the mastodon, both of large dimensions. Various families, and even genera, of quadrupeds now existing were associated with these extraordinary creatures, though of extinct species, such as the elephant, rhinoceros, hippopotamus, tapir, horse, bear, wolf, hyæna, weasel, beaver, ox, buffalo, deer, &c.; and also marine mammalia, as dolphins, sea-calves, walruses, and lamantines. Indeed, in the constant increase of animal life manifested throughout the whole of the tertiary strata, the forms approach nearer to living species as their remains lie high in the series.

In the older Pleiocene period some of the large amphibious quadrupeds, and other genera of mammalia of the earlier tertiary periods, appear no more; but there were the mastodon, and the elephas primogenius, or mammoth, some species of which, of prodigious size, were associated with numerous quadrupeds of existing genera, but lost species. Extinct species of almost all the quadrupeds now alive seem to have inhabited the earth at that time; their bones have been discovered in caverns; they were imbedded in the breccias and in most of the strata of that epoch—as the hippopotamus, rhinoceros, elephant, horse, bear, wolf, water-rat, hyæna, and various birds. It is remarkable that in the caverns of Australia the fossil bones all belong to extinct species of gigantic kangaroos and wombats, animals belonging to the marsupial family, which are so peculiarly the inhabitants of that country at the present day, but of diminished size. The newer Pleiocene strata show that the same analogy existed between the extinct and recent mammalia of South America, which, like their living congeners, as far as we know, belonged to that continent alone; for the fossil remains, quite different from those in the old world, are of animals of the same genera with the sloths, anteaters, and armadilloes, which now inhabit that country, but of vastly superior size and different species. The megatherium and equus curvidens, or extinct horse, had so vast a range in America, that, while Mr. Lyell collected their bones in Georgia, in 3° N. latitude, Mr. Darwin brought them from the corresponding latitude in South America. The equus curvidens differed as much from the living horse as the quagga or zebra does, and the European fossil horse is also a distinct and lost animal.

The greater part of the land in the northern hemisphere was elevated above the deep during the tertiary period, and such lands as already existed acquired additional height; consequently the climate, which had previously been tropical, became gradually colder, for an increase of land, which raises the temperature between the tropics, has exactly the contrary effect in higher latitudes. Hence excessive cold prevailed during the latter part of the Pleiocene period, and a great part of the European continent was covered by an ocean full of floating ice, not unlike that experienced at this day off the north-eastern coast of America.

During the latter part of the Pleiocene period, however, the bed of that glacial ocean rose partially, and after many vicissitudes the European continent assumed nearly the form and climate it now has. There is every reason to believe that the glacial sea extended also over great portions of the arctic lands of Asia and America. Old forms of animal and vegetable life were destroyed by these alterations in the surface of the earth and the consequent change of temperature; and when in the progress of the Pleiocene period the mountain-tops appeared as islands above the water, they were clothed with the flora and peopled by the animals they still retain; and new forms were added as the land rose and became dry and fitted to receive and maintain the races of beings now alive, all of which had possession of the earth for ages prior to the historical or human period. Some of the extinct animals had long resisted the great vicissitudes of the times; of these the mammoth, or elephas primogenius, whose fossil remains are found all over Europe, Asia, and America, but especially in the gelid soil of Siberia, alone outlived its associates, the last remnant of a former world. In two or three instances this animal has been discovered entire, entombed in frozen mud, with its hair and its flesh so fresh that wolves and dogs fed upon it. It has been supposed that, as the Siberian rivers flow for hundreds of miles from the southern part of the country to the Arctic Ocean, these elephants might have been drowned by floods while browsing in the milder regions, and that their bodies were carried down by the rivers and imbedded in mud, and frozen before they had time to decay. Although the congeners of this animal are now the inhabitants of the torrid zone, they may have been able to endure the cold of a Siberian winter. Baron Cuvier found that this animal differed as much from the living elephant as a horse does from an ass. The supply of food in summer was probably sufficient, since the quantity requisite for the maintenance of the larger animals is by no means in proportion to their bulk, and it may have migrated to a more genial climate in the cold months.

Shell-fish seem to have been more able to endure all the great geological changes than any of their organic associates; they show a constant approximation to modern species during the progress of the tertiary periods. The whole of these strata contain enormous quantities of shells of extinct species; in the oldest, three and a half per cent. of the shells are identical with some now existing, while on the uppermost strata of this geological period there are not less than from ninety to ninety-five in a hundred identical with those now alive.

Of all the fossil fishes from the silurian strata to the end of the tertiary, not one is specifically the same with living forms, except the Mallotus villosus, or captan, of the salmon family, and perhaps a few others of the most recent of these periods. In the Eiocene strata one-third belong to extinct genera.

Under the vegetable mould in every country there is a stratum of loose sand, gravel, and mud lying upon the subjacent rocks, often of great thickness, called alluvium, which in the high latitudes of North America and Europe is mixed with enormous fragments of rock, sometimes angular and sometimes rounded and waterworn, which have been transported hundreds of miles from their origin. It is there known as the Boulder formation, or Northern Drift, because, from the identity of the boulders with the rocks of the northern mountains, they evidently have come from them, and their size becomes less as the distance increases. In Russia there are blocks of great magnitude that have been carried eight hundred and even a thousand miles south-east from their origin in the Scandinavian range. There is every reason to believe that such masses, enormous as they are, have been transported by icebergs and deposited when the northern parts of the continents were covered by the glacial sea. The same process is now in progress in the high southern latitudes.

The last manifestation of creative power, with few exceptions, differs specifically from all that went before; the recent strata contain only the exuviæ of animals now living, often mixed with the bones and the works of man.

The thickness of the fossiliferous strata up to the end of the tertiary formation has been estimated at about seven or eight miles; so that the time requisite for their deposition must have been immense. Every river carries down mud, sand, or gravel to the sea; the Ganges brings more than 700,000 cubic feet of mud every hour, the Yellow River in China 2,000,000, and the Mississippi still more; yet, notwithstanding these great deposits, the Italian hydrographer, Manfredi, has estimated that, if the sediment of all the rivers on the globe were spread equally over the bottom of the ocean, it would require 1000 years to raise its bed one foot; so at that rate it would require 3,960,000 years to raise the bed of the ocean alone to a height nearly equal to the thickness of the fossiliferous strata, or seven miles and a half, not taking account of the waste of the coasts by the sea itself; but if the whole globe be considered instead of the bottom of the sea only, the time would be nearly four times as great, even supposing as much alluvium to be deposited uniformly both with regard to time and place, which it never is. Besides, in various places the strata have been more than once carried to the bottom of the ocean and again raised above its surface by subterranean fires after many ages, so that the whole period from the beginning of these primary fossiliferous strata to the present day must be great beyond calculation, and only bears comparison with the astronomical cycles, as might naturally be expected, the earth being without doubt of the same antiquity with the other bodies of the solar system. What then shall we say if the time be included which the granitic, metamorphic, and recent series occupied in forming? These great periods of time correspond wonderfully with the gradual increase of animal life and the successive creation and extinction of numberless orders of being, and with the incredible quantity of organic remains buried in the crust of the earth in every country on the face of the globe.

Every great geological change in the nature of the strata was accompanied by the introduction of a new race of beings, and the gradual extinction of those that had previously existed, their structure and habits being no longer fitted for the new circumstances in which these changes had placed them. The change, however, never was abrupt, except at the beginning of the tertiary strata; and it may be observed that, although the mammalia came last, there is no proof of progressive development, for animals and plants of high organization appeared among the earliest of their kind.

The geographical distribution of animated beings was much more extensive in the ancient seas and land than in later times. In very remote ages the same animal inhabited the most distant parts of the sea; the corallines built from the Equator to within ten or fifteen degrees of the Pole; and, previous to the formation of the carboniferous strata, there appears to have been even a greater uniformity in the vegetable than in the animal world, though New Holland had formed even then a peculiar district, supposing the coal in that country to be of the same epoch as in Europe and America; but as the strata became more varied, species were less widely diffused. Some of the saurians were inhabitants of both the Old and New World, while others lived in the latter only. In the tertiary periods the animals of Australia and America differed nearly as much from those of Europe as they do at the present day. The world was then, as now, divided into great physical regions, each inhabited by a peculiar race of animals; and even the different species of shell-fish of the same sea were confined to certain shores. Of 405 species of shell-fish which inhabited the Atlantic Ocean during the early and middle part of the tertiary period, only twelve were common to the American and European coasts. In fact, the divisions of the animal and vegetable creation into geographical districts had been in the latter periods contemporaneous with the rise of the land, each portion of which as it rose above the deep had been clothed with a vegetation and peopled with creatures suited to its position with regard to the equator, and to the existing circumstances of the globe; and the marine creatures had no doubt been divided into districts at the same periods, because the bed of the ocean had been subject to similar changes.

The quantity of fossil remains is so great that probably not a particle of matter exists on the surface of the earth that has not at some time formed part of a living creature. Since the commencement of animated existence, zoophytes have built coral reefs extending hundreds of miles, and mountains of limestone are full of their remains all over the globe. Mines of shells are worked to make lime; ranges of hills and rock, many hundred feet thick, are almost entirely composed of them, and they abound in every mountain-chain throughout the earth. The prodigious quantity of microscopic shells discovered by M. Ehrenberg is still more astonishing; shells not larger than a grain of sand form entire mountains: a great portion of the hills of Casciana in Tuscany consist of chambered shells so minute that Signor Saldani collected 10,454 of them from one ounce of stone. Chalk is often almost entirely composed of them. Tripoli, a fine powder long in use for polishing metals, is almost entirely composed of shells; the polishing property is owing to their siliceous coats; and there are even hills of great extent consisting of this substance, the débris of an infinite variety of microscopic insects.

The facility with which many slates and clays are split is owing, in some instances, to layers of minute shells. Fossil fish are found in all parts of the world, and in all the fossiliferous strata, with the exception of some of the lowest, but each great geological period had species of fish peculiar to itself.

The remains of the great saurians are innumerable; those of extinct quadrupeds are very numerous; but there is no circumstance in the whole science of fossil geology more remarkable than the inexhaustible multitudes of fossil elephants that are found in Siberia. Their tusks have been an object of traffic in ivory for centuries, and in some places they have been in such prodigious quantities, that the ground is tainted with the smell of animal matter. Their huge skeletons are found from the borders of Europe through all northern Asia to its extremest point, and from the foot of the Altaï mountains to the shores of the Frozen Ocean, a surface equal in extent to the whole of Europe. Some islands in the Arctic Sea are composed almost entirely of their remains, mixed with the bones of various other animals of living genera, but extinct species.

Equally wonderful is the quantity of fossil plants that still remain, if it be considered that from the frail nature of many vegetable substances multitudes must have perished without leaving a trace behind. The vegetation that covered the terrestrial part of the globe previous to the formation of the carboniferous strata had far surpassed in exuberance the rankest tropical jungles. There are many coal-measures of great extent in various parts of the earth, especially in North America, where that of Pittsburg occupies an area of about fourteen thousand square miles; and that in the Illinois is not much inferior to the area of all England.

As coal is entirely a vegetable substance, some idea may be formed of the richness of the ancient flora; in latter times it was less exuberant, and never has again been so luxuriant, probably on account of the decrease of temperature during the deposition of the tertiary strata, and in the glacial period which immediately preceded the creation of the present tribes of plants and animals. Even after their introduction the temperature must have been very low, but by subsequent changes in the distribution of the sea and land the cold was gradually mitigated, till at last the climate of the northern hemisphere became what it is now.

Such is the marvellous history laid open to us on the earth's surface. Surely it is not the heavens only that declare the glory of God,—the earth also proclaims His handiwork!

CHAPTER II.

FORM OF THE GREAT CONTINENT.—THE HIGH LANDS OF THE GREAT CONTINENT:—THE ATLAS, SPANISH, FRENCH, AND GERMAN MOUNTAINS—THE ALPS, BALKAN, AND APENNINES.

At the end of the Tertiary period the earth was much in the same state that it is at present with regard to the distribution of land and water. The preponderance of land in the northern hemisphere indicates a prodigious accumulation of internal energy under these latitudes at a very remote geological period. The forces that raised the two great continents above the deep, when viewed on a wide scale, must evidently have acted at right angles to one another, nearly parallel to the equator in the old continent, and in the direction of the meridian in the new; yet the structure of the opposite coasts of the Atlantic points at some connexion between the two.

The tendency of the land to assume a peninsular form is very remarkable; and it is still more so that almost all the peninsulas tend to the south, while to the north, with a very few exceptions, the two great continents terminate in a very broken line, and as they sink under the Icy Ocean, the tops of their high lands and mountains rise above the waves and stud the coast with innumerable snow-clad rocks and islands. Eastern Asia is evidently continued in a subaqueous continent from the Indian Ocean across the Pacific nearly to the west coast of America, of which New Holland, the Indian Archipelago, the islands of the Asiatic coast and of Oceania, are the great table-lands and summits of its mountain-chains.

Of the Polar lands little is known. Greenland probably is part of a continent, the domain of perpetual snow; and the recent discovery of so extensive a mass of high volcanic land near the South Pole is an important event in the history of physical science, though the stern severity of the climate must for ever render it unfit for the abode of animated beings, or even for the support of vegetable life. It seems to form a counterpoise to the preponderance of dry land in the northern hemisphere. There is something sublime in the contemplation of these lofty and unapproachable regions—the awful realm of ever-during ice and perpetual fire, whose year consists of one day and one night. The strange and terrible symmetry in the nature of the lands within the Polar circles, whose limits are to us a blank, where the antagonist principles of cold and heat meet in their utmost intensity, fills the mind with that awe which arises from the idea of the unknown and the indefinite.

The mountains, from their rude and shattered condition, bear testimony to repeated violent convulsions similar to modern earthquakes; while the high table-lands, and that succession of terraces by which the continents sink down from their mountain-ranges to the plains, to the ocean, and even below it, show also that the land must have been heaved up occasionally by slow and gentle pressure, such as appears now to be gradually elevating the coast of Scandinavia and many other parts of the earth. The periods in which these majestic operations were effected must have been incalculable, since the dry land occupies an area of nearly thirty-eight millions of square miles.

The division of the land is very unequal: the great continent has an area of about twenty-four millions of square miles, while the extent of America is about eleven millions, and that of Australia, with its islands, scarcely three; Africa is more than three times the size of Europe, and Asia is more than four times as large.

The peninsular form of the continents adds greatly to the extent of their coasts, of such importance to civilization and commerce. All the shores of Europe are deeply indented and penetrated by the Atlantic Ocean, which has formed a number of inland seas of great magnitude, so that it has a greater line of maritime coast compared with its size than any other quarter of the world. The extent of coast from the Straits of Waigatz in the Polar Ocean to the Strait of Caffa at the entrance of the Sea of Azoff, is about seventeen thousand miles. The coast of Asia has been much worn by currents, and possibly also by the action of the ocean occasioned by the rotation of the earth from west to east. On the south and east especially it is indented by large seas, bays, and gulfs; and the eastern shores are rugged, and encompassed by chains of islands which render navigation dangerous. Its maritime coast is about thirty-three thousand miles in length.

The coast of Africa, sixteen thousand miles long, is very entire, except perhaps at the Gulf of Guinea and in the Mediterranean. The shores of North America have probably been much altered by the equatorial current and the gulf-stream. There cannot be a doubt that these currents, combined with volcanic action, have hollowed out the Gulf of Mexico, and separated the Antilles and Bahama Islands from the continent. The coast is less broken on the west, but in the Icy Ocean there is a labyrinth of gulfs, bays, and creeks. The shores of South America, on both sides, are very entire, except towards Cape Horn and Southern Chili, where the tremendous surge and currents of the ocean in those high latitudes have eaten into the mountains, and produced endless irregularities and fiords, which run far into the land. The whole continent of America has a sea-coast of thirty-one thousand miles. Thus it appears that the ratio of the number of linear miles in the coast-line to that of square miles in the extent of surface, in each of these great portions of the globe, is 164 for Europe, 376 for Asia, 530 for Africa, and 359 for America. Hence the proportion is most favourable to Europe with regard to civilization and commerce; America comes next, then Asia, and last of all Africa, which has every natural obstacle to contend with, from the extent and nature of its coasts, the desert character of the country, and the unwholesomeness of its climate, on the Atlantic coast at least.

The continents had been raised from the deep by a powerful effort of the internal forces acting under widely-extended regions, and the stratified crust of the earth either remained level, rose in undulations, or sank into cavities, according to its intensity. Some thinner portion of the earth's surface, giving way to the internal forces, had been rent into deep fissures, and the mountain masses had been raised by violent concussions, perceptible in the convulsed state of their strata. The centres of maximum energy are marked by the pyrogenous rocks which generally form the nucleus or axis of the mountain masses, on whose flanks the stratified rocks are tilted at all angles to the horizon, whence declining on every side they sink to various depths or stretch to various distances on the plains. Enormous as the mountain-chains and table-lands are, and prodigious as the forces that elevated them, they bear a very small proportion to the mass of the level continents and to the vast power which raised them even to their inferior altitude. Both the high and the low lands had been elevated at successive periods; some of the very highest mountain-chains are but of recent geological date, and some chains that are now far inland once stood up as islands above the ocean, while marine strata filled their cavities and formed round their bases. The influence of mountain-chains on the extent and form of the continents is beyond a doubt.

Notwithstanding the various circumstances of their elevation, there is everywhere a certain regularity of form in mountain masses, however unsymmetrical they may appear at first, and rocks of the same kind have identical characters in every quarter of the globe. Plants and animals vary with climate, but a granite mountain has the same peculiarities in the southern as in the northern hemisphere, at the equator as near the poles. Single mountains, insulated on plains, are rare, except where they are volcanic; they generally appear in groups intersected by valleys in every direction, and more frequently in extensive chains symmetrically arranged in a series of parallel ridges, separated by narrow longitudinal valleys, the highest and most rugged of which occupy the centre: when the chain is broad and of the first order in point of magnitude, peak after peak arise in endless succession. The lateral ridges and valleys are constantly of less elevation, and are less bold, in proportion to their distance from the central mass, till at last the most remote ridges sink down into gentle undulations. Extensive and lofty branches diverge from the principal chains at various angles, and stretch far into the plains. They are often as high as the chains from which they spring, and it happens not unfrequently that these branches are united by transverse ridges, so that the country is often widely covered by a network of mountains, and, at the point where these offsets diverge, there is frequently a knot of mountains spreading over hundreds of square miles. The circumstances of elevation are not the only causes of that variety observed in the summits of mountain-chains; a very minute difference in the composition and internal structure of a rock has great influence upon its general form, and on the degree and manner in which it is worn by the weather.

One side of a mountain-range is usually more precipitous than the other, but there is nothing in which the imagination misleads the judgment more than in estimating the steepness of a declivity. In the whole range of the Alps there is not a single rock which has 1600 feet of perpendicular height, or a vertical slope of 90°. The declivity of Mont Blanc towards the Allée Blanche, precipitous as it seems, does not amount to 45°; and the mean inclination of the Peak of Teneriffe, according to Baron Humboldt, is only 12° 30′. The Silla of Caraccas, which rises precipitously from the Caribbean Sea, at an angle of 53° 28′, to the height of between six and seven thousand feet, is a majestic instance of the nearest approach to perpendicularity of any great height yet known.

Immediately connected with the mountains are the high table-lands which form so conspicuous a feature in the Asiatic and American continents. These perpetual storehouses of the waters send their streams to refresh the plains, and to afford a highway between the nations. Table-lands of less elevation, sinking in terraces of lower and lower level, constitute the links between the high ground and the low, the mountains and the plains, and thus maintain the continuity of the land. They frequently are of the richest soil, and enjoy the most genial climate, affording a delightful and picturesque abode to man, though the plains are his principal dwelling. Sloping imperceptibly from the base of the inferior table-lands, or from the last undulations of the mountains to the ocean, they carry off the superfluous waters. Fruitfulness and sterility vary their aspect; immense tracts of the richest soil are favoured by climate and hardly require culture; a greater portion is only rendered productive by hard labour, compelling man to fulfil his destiny; while vast regions are doomed to perpetual barrenness, never gladdened by a shower.

The form of the great continent has been determined by an immense zone of mountains and table-lands, lying between the 30th and 40th or 45th parallels of north latitude, which stretches across it from W.S.W. to E.N.E., from the coasts of Barbary and Portugal on the Atlantic Ocean to the farthest extremity of Asia at Behring's Straits in the North Pacific. North of this lies an enormous plain, extending almost from the Pyrenees to the utmost part of Asia, the greater portion of which is a dead level, or low undulations, uninterrupted, except by the Scandinavian and British system on the north, and the Ural chain, which is of small elevation. The low lands south of the mountainous zone are much indented by the ocean, and of the most diversified aspect. By much the greater part of the flat country lying between the China Sea and the river Indus is of the most exuberant fertility, while that between the Persian Gulf and the foot of the Atlas is, with some happy exceptions, one of the most desolate tracts on the earth. These southern lowlands, too, are broken by a few mountain systems of considerable extent and height.

The Atlas and Spanish mountains form the western extremity of that great zone of high lands that girds the old continent almost throughout its extent. These two mountain systems were certainly at one time united; and, from their geological formation, and also the parallelism of their mountain-chains, they must have been elevated by forces acting in the same direction,—now, indeed, the Straits of Gibraltar, a sea-filled chasm of unfathomable depth, divides them.

A very elevated and continuous mountainous region extends in a broad belt along the north-west of Africa, from the promontory of Gher on the Atlantic to the Gulf of Sidra on the Mediterranean, enclosing all the high lands of Morocco, Algiers, and Tunis. It is bounded by the Atlantic and Mediterranean, and insulated from the rest of Africa by the Sahara desert.

This mountain system consists of three parts. The chain of the Greater Atlas, which is farthest inland, extends from Souse near the Atlantic to the Lesser Syrte, and in Morocco forms a mountain-knot 15,000 feet high, perpetually covered with snow.

The Lesser Atlas begins at Cape Kotes opposite to Gibraltar, and keeps parallel to the Mediterranean till it attains the Gharian range in Tripoli, the last and lowest of the Little Atlas, which runs due east in a uniformly diminishing line till it vanishes in the plains of the Great Syrte. That long, rugged, but lower chain of parallel ridges and groups, which forms the bold coasts of the Straits of Gibraltar and the Mediterranean, is only a portion of the Lesser Atlas, which rises above it majestically, covered with snow. The flanks of the mountains are generally covered with forests, but their summit is one uninterrupted line of bare inaccessible rocks, and they are rent by fissures frequently not more than a few feet wide,—a peculiar feature of the whole system.

The Middle Atlas, lying between the two great chains, consists of a table-land, rich in valleys and rivers, which rises in successive terraces to the foot of the Greater Atlas, separated by ranges of hills parallel to it. This wide and extensive region has a delightful climate, abounds in magnificent forests, and the valleys are full of vitality. The crest of the Atlas is of granite and crystalline strata; their flanks and lower ranges are sandstone and limestone, on which the tertiary strata rest.

The Spanish peninsula consists chiefly of a table-land traversed by parallel ranges of mountains, and surrounded by the sea, except where it is separated from France by the Pyrenees, which extend from the Mediterranean to the Bay of Biscay, but are continued by the Cantabrian chain to Cape Finisterre on the Atlantic.

The Pyrenean chain is of moderate height at its extremities, but its summit maintains a waving line whose mean altitude is 7000 feet; it rises to a greater height on the east; its highest point is the Pic du Midi, 11,000 feet above the sea. The snow lies deep on these mountains during the greater part of the year, and is perpetual on the highest parts; but the glaciers, which are chiefly on the northern side, are neither so numerous nor so large as in the Alps.

The greatest breadth of this range is about sixty miles, and its length two hundred and seventy. It is so steep on the French side, so rugged, and so notched, that from the plains below its summits look like the teeth of a saw, whence the term Sierra has been appropriated to mountains of this form. On the Spanish side, gigantic sloping offsets, separated by deep precipitous valleys, penetrate to the banks of the Ebro. All the Spanish mountains are torn by deep crevices, the beds of torrents and rivers.

The interior of Spain is a table-land, with an area of 93,000 square miles, nearly equal to half of the peninsula. It dips to the Atlantic from its western side, where its altitude is about 2500 feet. There it is bounded by the Iberian mountains, which begin at the point where the Pyrenees take the name of the Cantabrian chain, and run in a tortuous south-easterly direction through all Spain, constituting the western boundary of Valencia and Murcia, and sending many branches through those provinces to the Mediterranean. Its most elevated point is the Sierra Urbian, 7272 feet high.

Four nearly parallel ranges of mountains originate in this limiting chain, running from N.E. to S.W. diagonally across the peninsula to the Atlantic. Of these the high Castilian mountains and the Sierra di Toledo cross the table-land; the Sierra Morena, so called from the dingy colour of its forests of Hermes oak, on the southern edge; and, lastly, the Sierra Nevada, though only a hundred miles long and fifty broad, the finest range of mountains in Europe after the Alps, traverses the plains of Andalusia and Grenada. The table-land is monotonous and bare of trees; the plains of Old Castile are as naked as the steppes of Siberia, and uncultivated except along the banks of the rivers. Corn and wine are produced in abundance on the wide plains of New Castile and Estremadura; other places serve for pasture. The table-land becomes more fertile as it descends towards Portugal, which is altogether more productive than Spain, though the maritime provinces of the latter on the Mediterranean are luxuriant and beautiful with a semi-tropical vegetation.

Granite, crystalline strata, and primary fossiliferous rocks prevail chiefly in the Spanish mountains, and give them their peculiar bold serrated aspect. The tracts between the parallel ranges through which the great Spanish rivers flow to the Atlantic appear to have been at one time the basins of lakes.

The mass of the high land is continued through the south of France, at a much lower elevation, by chains of hills and table-lands, the most remarkable of which are the Montagnes Noires, and the great platform of Auvergne, once the theatre of violent volcanic action. It continued from the beginning to the middle of the tertiary period, so that there are craters of various ages and perfect form: some of the highest, as the Puy de Dôme, 5000 feet high, are trachytic craters of elevation; Mont Dore, 6200 feet high, is probably the most elevated. These volcanic mountains of Auvergne, and the Cevennes, above 6000 feet high, are the most remarkable of the French system; the offsets of the latter reach the right bank of the Rhone and the Jura mountains of the Alpine range. In fact, the French mountains are the link between the more elevated masses of western and eastern Europe.

The eastern and highest part of the European portion of the mountain-zone begins to rise above the low lands about the 52nd parallel of north latitude, ascending by terraces, groups, and chains of mountains, through six or seven degrees of latitude, till it reaches its highest point in the great range of the Alps and Balkan. The descent on the south side of this lofty mass is much more rapid and abrupt, and the immediate offsets from the Alps shorter; but, taking a very general view, the Apennines and mountains of northern Sicily, those of Greece and the southern part of Turkey in Europe, with all the islands of the adjacent coasts, are but outlying members of the general protuberance.

The principal chain of the Hyrcanian mountains, the Sudetes, and the Carpathian mountains, form the northern boundary of these high lands: the first, consisting of three parallel ridges, extends from the right bank of the Rhine to the centre of Germany, about 51° or 52° of N. lat., with a mean breadth of about a hundred miles, and terminates in the knot of the Fichtelberge, covering an area of 9000 square miles, on the confines of Bavaria and Bohemia. The

Sudetes begin on the east of this group, and, after a circuit of three hundred miles round Bohemia, terminate at the small elevated plain of the Upper Oder, which connects them with the Carpathian mountains. No part of these limiting ranges attains the height of 5000 feet, except the Carpathians, some of which are very high. They consist of mountain groups, united by elevated plains, rather than of a single chain: the Tatra mountains, bisected by the 20th meridian, is their loftiest point. This range is high also in Transylvania, before it reaches the Danube, which divides it from a secondary branch of the Balkan. Spurs decline in undulations from these limiting chains on the great northern plain, and the country to the south, intervening between them and the Alps, is covered with an intricate network of mountains and plains of moderate elevation.

The higher Alps, which form the western crest of the elevated zone, begin at the Capo della Melle, on the Gulf of Genoa, and bend round by the west and north to Mont Blanc; then turning E.N.E. they run through the Grisons and Tyrol to the Great Glockner in 40° 7′ N. lat. and 12° 43′ E. long., where the higher Alps terminate a course 420 miles long. All this chain is lofty; much of it is above the line of perpetual congelation, but the most elevated part lies between the Col de la Seigne, on the west shoulder of Mont Blanc, and the Simplon. The highest mountains in Europe are comprised within this space, not more than sixty miles long, where Mont Blanc, the highest of all, has an absolute elevation of 15,730 feet. The central ridge of the higher Alps is jagged with peaks, pyramids, and needles of bare and almost perpendicular rock, rising from fields of perpetual snow and rivers of ice to an elevation of 14,000 feet. Many parallel chains and groups, alike rugged and snowy, press on the principal crest, and send their flanks far into the lower grounds. Innumerable secondary branches, hardly lower than the main crest, diverge from it in various directions ; of these the chain of the Bernese Alps is the highest and most extensive. It breaks off at St. Gothard, in a line parallel to the principal chain, separates the Valais from the canton of Bern, and with its ramifications forms one of the most remarkable groups of mountain scenery in Europe. Its endless maze of sharp ridges and bare peaks, mixed with gigantic masses of pure snow fading coldly serene into the blue horizon, present a scene of sublime quiet and repose, unbroken but by the avalanche or the thunder.

At the Great Glockner, the range of the Alps, hitherto undivided, splits into two branches, the Noric and Carnic Alps : the latter is the continuation of the chief stem. Never rising to the height of perpetual snow, it separates the Tyrol and Upper Carinthia from the Venetian States, and, taking the name of the Julian Alps at Mont Terglou, 9380 feet above the sea, runs east till it joins the eastern Alps or Balkan, under the 18th meridian. Offsets from this chain cover all the neighbouring countries.

It is difficult to estimate the width of the Alpine

chain; that of the higher Alps is about a hundred miles; it increases to a hundred and fifty east of the Grisons, and amounts to two hundred between the 15th and 16th meridians, but is not more than eighty at its junction with the Balkan.

The Stelvio, 9174 feet above the sea, is the highest carriage-pass in these mountains. That of St. Gothard is the only one which goes directly over the crest of the Alps. Passes very rarely go over the summit of a mountain; they generally cross the water-shed, ascending by the valley of a torrent, and descending by a similar path on the other side.

The frequent occurrence of extensive deep lakes is a peculiar feature in European mountains, rarely to be met with in the Asiatic system, except in the Altaï, and on the elevated plains.

With the exception of the Jura, whose pastoral summit is about 3000 feet above the sea, there are no elevated table-lands in the Alps; the tabular form, so eminently characteristic of the Asiatic high lands, begins in the Balkan. The Oriental peninsula rises by degrees from the Danube to Bosnia and Upper Macedonia, which are some hundred feet above the sea; and the Balkan extends six hundred miles along this elevated mass, from the Julian Alps to Cape Eminek on the Black Sea. It begins by a table-land seventy miles long, traversed by low hills, ending towards Albania and Myritida in a limestone wall from six to seven thousand feet high. Rugged mountains, all but impassable, succeed to this, in which the domes and needles of the Schandach, or ancient Scamus, are covered with perpetual snow. Another table-land follows, whose marshy surface is bounded by mural precipices ending at Mount Arbelus, 9000 feet high, near the town of Sophia. There the Hemus, or Balkan properly so called, begins, and runs in parallel ridges, separated by fertile longitudinal valleys, to the Black Sea, dividing the plains between the Lower Danube and the Propontis into nearly equal parts. The central ridge rises at once in a wall 4000 feet high, passable in few places; and where there is no lateral ridge the precipices descend at once to the plains.

The Balkan is everywhere rent by terrific fissures across the chains and table-lands, so deep and narrow that daylight is almost excluded. These chasms afford the safest passes across the range; the others, along the faces of the precipices, are frightful.

The Mediterranean is the southern boundary of the elevated zone of Eastern Europe, whose last offsets rise in rocky islands along the coasts. The crystalline mountains of Sardinia and Corsica are outlying members of the Maritime Alps, while shorter offsets end in the plains of Lombardy, forming the magnificent scenery of the Italian lakes. Even the Apennines, whose elevation has given its form to the peninsula of Italy, is but a secondary, on a greater scale, to the broad central band, as well as the mountains and high land in the north of Sicily, which form the continuation of the Calabrian chain.

The Apennines, beginning at the Maritime Alps,

enclose the Gulf of Genoa, and run through the centre of Italy in parallel ranges to the middle of Calabria, where they split into two branches, one of which goes to Capo de Leuca on the Gulf of Torento, the other to Cape Spartivento in the Straits of Messina. The whole length is about eight hundred miles. None of the Apennines come within the line of perpetual snow, though it lies nine months in the year on the Gran Sasso d'Italia, 9521 feet high in Abruzza Ulteriore.

Offsets from the Julian and Eastern Alps render Dalmatia and Albania perhaps the most rugged tract in Europe; and the Pindus, which forms the water-shed of Greece, diverges from the latter chain, and, running south two hundred miles, separates Albania from Macedonia and Thessaly.

Greece is a country of mountains, and, although none are perpetually covered with snow, it lies nine months on several of their summits. The chains terminate in strongly projecting headlands, which reach far into the sea, and reappear in the numerous islands and rocks which stud that deeply indented coast. The Grecian mountains, like the Balkan, are torn by transverse fractures. The celebrated Pass of Thermopylæ, the defile of Blatamana, and the Gulf of Salonica are examples. The Adriatic, the Dardanelles, and the Sea of Marmora limit the secondaries of the southern part of the Balkan.

The valleys in the Alps are long and narrow; those among the mountains of Turkey in Europe and Greece are mostly caldron-shaped hollows, often

enclosed by mural rocks. Many of these cavities of great size lie along the foot of the Balkan. In the Morea they are so encompassed by mountains that the water has no escape but through the porous soil. They consist of tertiary strata, which had formed the bottom of lakes. Caldron-shaped valleys occur in most volcanic countries, as Sicily, Italy, and central France.

The table-lands which constitute the tops of mountains or of mountain-chains are of a different character from those terraces by which the high lands slope to the low. The former are on a small scale in Europe, and of a forbidding aspect, with the exception of the Jura, which is pastoral; whereas the latter are almost always habitable and cultivated. The mass of high land in South-Eastern Europe shelves on the north to the great plain of Bavaria, 3000 feet high; Bohemia, which slopes from 1500 to 900, and Hungary, from 4000 above the sea to 300. The descent on the south of the Alps is six or seven times more rapid, because the distance from the axis of the chain is shorter.

It is scarcely possible to estimate the quantity of ice in the Alps; it is said, however, that, independent of the glaciers in the Grisons, there are 1500 square miles of ice in the Alpine range, from eighty to six hundred feet thick. Some glaciers have been permanent and stationary in the Alps time immemorial, while others now occupy ground formerly bearing corn or covered with trees, which the irresistible force of the ice has swept away. These ice

rivers, formed on the snow-clad summits of the mountains, fill the hollows and high valleys, hang on the declivities, or descend by their weight through the transverse valleys to the plains, where they are cut short by the increased temperature, and deposit those accumulations of rocks and rubbish, called moraines, which had fallen upon them from the heights above. In the Alps the glaciers move at the rate of from twelve to twenty-five feet annually, and, as in rivers, the motion is most rapid in the centre. They advance or retreat according to the mildness or severity of the season, but they have been subject to cycles of unknown duration. From the moraines, as well as the striæ engraven on the rocks over which they have passed, M. Agassiz has ascertained that the valley of Chamouni was at one time occupied by a glacier that had moved towards the Col di Balme. A moraine 2000 feet above the Rhone at St. Maurice shows that at a remote period glaciers had covered Switzerland to the height of 2155 feet above the Lake of Geneva.

Their increase is now limited by various circumstances—as the mean temperature of the earth, which is always above the freezing-point in those latitudes; excessive evaporation; and blasts of hot air, which occur at all heights, in the night as well as in the day, from some unknown cause. They are not peculiar to the Alps, but have been observed also on the glaciers of the Andes. Besides, the greater the quantity of snow in the higher Alps, the lower is the glacier forced into the plains.

Granite no doubt forms the base of the mountain system of Eastern Europe, though it more rarely comes into view than might have been expected. Crystalline schists of various kinds are enormously developed, and generally form the most elevated pinnacles of the Alpine crest and its offsets; but the secondary fossiliferous strata constitute the chief mass, and often rise to the highest summits; indeed, secondary limestones occupy a great portion of the high land of Eastern Europe. Calcareous rocks form two great mountain-zones on each side of the central chain of the Alps, and rise occasionally to altitudes of ten or twelve thousand feet. They constitute the central range of the Apennines, and fill the greater part of Sicily. They are extensively developed in Turkey in Europe, where the plateau of Bosnia with its high lands on the south, part of Macedonia, and Albania with its islands, are principally composed of them. Tertiary strata, of great thickness, rest on the flanks of the Alps, and rise in some places to a height of five thousand feet. Zones of the older Pleiocene period flank the Apennines on each side, filled with organic remains; and half of Sicily is covered with the newer Pleiocene strata.

From numerous dislocations in the strata, the Alps appear to have been heaved up by many violent and repeated convulsions, separated by intervals of repose, and different parts of the chain have been raised at different times; for example, the Maritime Alps and the south-western part of the Jura mountains were raised previous to the formation of the chalk: but the tertiary period appears to have been that of the greatest commotions; for nearly two-thirds of the lands of Europe have risen since the beginning of that epoch, and those that existed then acquired additional height, though some sank below their original level. During that time the Alps acquired an additional elevation of between two and three thousand feet: Mont Blanc then reached its present altitude; the Apennines rose one or two thousand feet higher; and the Carpathians seem to have gained an accession of height about the same period. That part of the Alpine chain lying between Mont Blanc and Vienna is said to have acquired its last accession of height since the seas were inhabited by the existing species of animals.

CHAPTER III.

THE HIGH LANDS OF THE GREAT CONTINENT, *continued:*—THE CAUCASUS—THE WESTERN ASIATIC TABLE-LAND AND ITS MOUNTAINS.

The Dardanelles and the Sea of Marmora form but a small break in the mighty girdle of the old continent, which again appears in immense table-lands passing through the centre of Asia, of such magnitude that they occupy nearly two-fifths of the continent. Here everything is on a much grander scale than in Europe; the table-lands rise above the mean height of the European mountains, and the mountains themselves that gird and traverse them surpass those of every other country in altitude. The most barren deserts are here to be met with, as well as the most luxuriant productions of animal and vegetable life. The earliest records of the human race are found in this cradle of civilization, and monuments still remain which show the skill and power of those nations which have passed away, but whose moral influence is still visible in their descendants. Customs, manners, and even prejudices, carry us back to times beyond the record of history, or even of tradition; while the magnitude with which the natural world is here developed evinces the tremendous forces that must have been in action at epochs immeasurably anterior to the existence of man.

The gigantic mass of high land which extends for 6000 miles between the Mediterranean and the Pacific is 2000 miles broad at its eastern extremity, 700 to 1000 in the middle, and somewhat less at its western termination. Colossal mountains and elevated terraces form the edges of these lofty plains.

Between the 47th and 68th eastern meridians, where the low plains of Hindostan and Bucharia press upon the table-land and reduce its width to 700 or 1000 miles, it is divided into two parts by an enormous knot of mountains formed by the meeting of the Hindoo Coosh, the Himalaya, the Thsung-ling, and the transverse ranges of the Beloot Tagh, or Cloudy Mountains: these two parts differ in height, form, and magnitude.

The western portion, which is the table-land of Persia or plateau of Iran, is oblong, extending from the shores of Asia Minor to the Hindoo Coosh and the Solimaun range, which skirts the right bank of the Indus. It occupies an area of 1,700,000 square miles, generally about 4000 feet above the sea, and in some places 7000. The oriental plateau or table-land of Tibet, much the largest, has an area of 7,600,000 square miles, and a mean altitude of 14,000 feet, and in some parts of Tibet an absolute altitude of 17,000 feet.

As the table-lands extend from S.W. to N.E., so also do the principal mountain-chains, as well those which bound the high lands as those which traverse them, with the exception of the Beloot Tagh, or Bolor, and the Solimaun chains, which run from north to south. The first is the western limit of the oriental plateau, the other the boundary of the table-land of Persia.

The lofty range of the Caucasus, which extends 700 miles between the Black and Caspian Seas, is an outlying member of the Asiatic high lands. Offsets diverge like ribs from each side of the central crest, which penetrate the Russian steppes on one hand, and on the other cross the plains of Kara, or valley of the Kour and Rioni, and unite the Caucasus to the table-land. Some parts of these mountains are more than 15,000 feet high; the Elbrouz, on the western border of Georgia, is 17,796 feet. The central part of the chain is full of glaciers, and the limit of perpetual snow is at the altitude of 11,000 feet, which is higher than in any other chain, except the Himalaya.

Anatolia, the most western part of the table-land of Iran, 3000 feet above the sea, is traversed by short chains and broken groups of mountains, separated by fertile valleys which sink rapidly towards the archipelago and end in promontories and islands along the shores of Asia Minor, which is a country abounding in vast luxuriant but solitary plains, watered by broad rivers. Single mountains of volcanic formation are conspicuous objects on the table-land of Anatolia, which is rich in pasture, though much of the soil is saline and covered with lakes and marshes. A triple range of limestone mountains, 6000 or 7000 feet high, divided by narrow but beautiful valleys, is the limit of the Anatolian table-land along the shores of the Black

Sea. They are covered with forests to the height of 4500 feet, and broken by wooded glens, having a narrow coast, except near Trebizond, where it is broad and picturesque. The high land is bounded on the south by the serrated snowy range of the Taurus, which, beginning in Rhodes, Cos, and other islands in the Mediterranean, fills the south-western parts of Asia Minor with ramifications, and, after following the sinuosities of the iron-bound coast of Karamania in a single lofty range, extends at Samisat, where the Euphrates has pierced a way through this stony girdle.

About the 50th meridian the table-land is compressed to nearly half its width, and there the lofty mountainous regions of Armenia, Kourdistan, and Azerbijan tower higher and higher between the Black Sea, the Caspian, and the Gulf of Alexandretta in the Mediterranean. Here the cold treeless plains of Armenia, the earliest abode of man, 7000 feet above the sea, bear no traces of the garden of Eden; but Mount Ararat, on which the ark is said to have rested, stands a solitary majestic volcanic cone 17,260 feet above the sea, shrouded in perpetual snow. Though high and cold, the soil of Armenia is better than that of Anatolia, and is better cultivated. It shelves on the north in luxuriant and beautiful declivities to the low and undulating valley of Kara, south of the Caucasus; and on the other hand, the broad and lofty belt of the Kourdistan Mountains, rising abruptly in many parallel ranges from the plains of Mesopotamia, form its southern limit, and spread their ramifications wide over its surface. They are rent by deep ravines, and in many places are so rugged that communication between the villages is always difficult, and in winter impracticable from the depth of snow. The line of perpetual congelation is decided and even along their summit; their flanks are wooded, and the valleys populous and fertile.

A thousand square miles of Kourdistan is occupied by the brackish lake Van, which is seldom frozen, though 5467 feet above the sea and surrounded by lofty mountains.

The Persian mountains, of which the Elbrouz is the principal chain, extend along the northern brink of the plateau, from Armenia, almost parallel to the shores of the Caspian Sea, maintaining a considerable elevation up to the volcanic mountain Demavend, near Tehran, their culminating point, 14,600 feet high, which, though 90 miles inland, is a landmark to sailors on the Caspian. Elevated offsets of these mountains cover the volcanic table-land of Azerbijan, the fire country of Zoroaster, and one of the best provinces in Persia; there the Koh Savalan elevates its volcanic cone 12,000 feet. Beautiful plains, pure streams, and peaceful glades, interspersed with villages, lie among the mountains, and the Vale of Khosran Shah, a picture of sylvan beauty, is celebrated as one of the five paradises of Persian poetry. The vegetation at the foot of these mountains on the shores of the Caspian has all the exuberance of a tropical jungle. The Elbrouz loses its height to the east of Demavend, and then joins the mountains of Khorasan and the Parapamisan range, which appear to be chains of mountains when viewed from the low plains of Khorasan and Balkh, but on the table-land of Persia they merely form a broad hilly country of rich soil till they join the Hindoo Coosh.

The table-land of Iran is bounded, for a thousand miles along the Persian Gulf and Indian Ocean, by a mountainous belt of from three to seven parallel ranges, having an average width of 200 miles, and extending from the extremity of the Kourdistan Mountains to the mouth of the Indus. The Lasistán Mountains, which form the northern part of this belt, and bound the vast level plain of the Tigris, rise from it in a succession of high table-lands divided by very rugged mountains, the last ridge of which, mostly covered with snow, abuts on the table-land of Persia. Oaks clothe their flanks; the valleys are of generous soil, verdant and cultivated; and many rivers flow through them to swell the stream of the Tigris. Insulated hill forts, from 2000 to 5000 feet high, occur in this country, with flat cultivated tops some miles in extent, accessible only by ladders or holes cut in their precipitous sides. These countries are full of ancient inscriptions and remains of antiquity. The moisture decreases more and more south from Shiraz, and then the parallel ridges, repulsive in aspect and difficult to pass, are separated by arid longitudinal valleys, which ascend like steps from the narrow shores of the Persian Gulf to the table-land. The coasts of the gulf are burning-hot sandy solitudes, so completely barren that the country from Bassora to the Indus, a distance of 1200 miles, is a sterile waste. In the few favoured spots on the terraces where water occurs there is vegetation, and the beauty of these valleys is enhanced by surrounding sterility.

With the exception of Mazenderan, and the other provinces on the Caspian and in the Parapamisan range, Persia is arid, possessing few perennial springs, and not one great river; in fact, three-tenths of the country is desert, and the table-land is nearly a wide scene of desolation. A great salt desert occupies 27,000 square miles between Irak and Khorasan, of which the soil is stiff clay covered with efflorescence of common salt and nitre, often an inch thick, varied only by a few saline plants and patches of verdure in the hollows. This dreary waste joins the large sandy and equally dreary desert of Kerman. Kelat, the capital of Belochistan, is 7000 feet above the level of the sea, round which there is cultivation, but the greater part of that country is a lifeless plain, over which the brick-red sand is drifted by the north wind into ridges like the waves of the sea, often twelve feet high, without a vestige of vegetation. The blast of the desert, whose hot and pestilential breath is fatal to man and animals, renders these dismal sands impassable at certain seasons.

Barren lands or bleak downs prevail at the foot of the Lukee and Solimaun ranges of bare porphyry and sandstone, which skirt the eastern edge of the table-land and dip to the plains of the Indus. In Afghanistan there is cultivation chiefly on the banks of the streams that flow into Lake Zorah, but vitality returns towards the north-east. The plains and valleys among the offsets from the Hindoo Coosh are of surpassing loveliness, and combine the richest peaceful beauty with the majesty of the snow-capped mountains.

CHAPTER IV.

THE HIGH LANDS OF THE GREAT CONTINENT, *continued*:—THE ORIENTAL TABLE-LAND AND ITS MOUNTAINS.

THE oriental plateau, or table-land of Tibet, is an irregular four-sided mass stretching from S.W. to N.E., enclosed and traversed by the highest mountains in the world. It is separated from the table-land of Persia by the Hindoo Coosh, a branch of the Himalaya, which occupies the terrestrial isthmus between the low lands of Hindostan and Bucharia.

The cold dreary plateau of Tibet is separated on the south from the glowing luxuriant plains of Hindostan by the Himalaya, which extends 2800 miles from the western extremity of the Hindoo Coosh in Cabulistan to the Gulf of Tonkin in China. The chain of the Altaï, to the north, 4500 miles long, divides the table-land from the deserts of Asiatic Siberia, and, stretching to the sea at Okhotzk under various names, it bends to the N.N.E., and terminates at Behring's Straits, the utmost extremity of Asia. The table-land terminates in the east, partly in the long Chinese chain of the Khing-Khan and Inshan Mountains, which stretch from the Altaï range to the great bend in the Yellow River in China, and farther south by the nameless and almost unknown magnificent mountains in the western provinces of the Chinese empire. On the west the table-land has its limits in the Beloot Tagh, or Cloudy Mountains, the Tartash Tagh of the natives, a transverse range, which leaves the Hindoo Coosh nearly at a right angle about the 72nd degree of E. longitude, and, pursuing a northerly direction, is supposed to unite the latter chain to that of the Altaï; its offsets, at least, extend widely in that direction. It forms magnificent mountain-knots with the diagonal chains of the table-land, and is the water-shed between Independent and Chinese Tourkistan, or Tartary. It descends in a succession of tiers or terraces through the countries of Bokhara and Balkh to the deep cavity in which the Caspian Sea and the Sea of Azoff lie, and forms, with the Paralasa, the Solimaun range, and the Ural, a singular exception to the general parallelism of Asiatic mountains. Two narrow difficult passes lead over the Beloot Tagh from the low plains of Bucharia and Independent Tourkistan to Kashgar and Yarkand, on the table-land in Chinese Tartary.

The table-land itself is crossed diagonally from west to east by two great chains of mountains. The Kuen-leun, or Chinese range, begins about 35° 30′ N. lat. at the mountain-knot formed by the Hindoo Coosh and Himalaya, and, running eastward, it terminates south of the Gulf of Petcheli, and covers a great part of the western provinces of China with its branches. The Thian-shan, or Celestial Mountains, lie more to the north; they begin at the Beloot Tagh, and, running along the 42nd parallel, sink to the desert of the Great Gobi, about the centre of the plateau, but, rising again, they end in various branches in China. The latter chain is exceedingly volcanic, and, though so far inland, pours forth lava, and exhibits all the other phenomena of volcanic districts.

Tibet is enclosed between the Himalaya and the Kuen-leun; Tungut, or Chinese Tartary, lies between the latter chain and the Celestial Mountains, and Zungary, or Mongolia, between the Celestial range and the Altaï. The Himalaya and Altaï ranges diverge in their easterly courses so that the table-land, which is only from 700 to 1000 miles wide at its western extremity, is 2000 between the Chinese province of Yunnan and the country of the Mantshu Tonguses.

Of all these vast chains of mountains the Himalaya, and its principal branch the Hindoo Coosh, are best known; though even of these a great part has never been explored, on account of their enormous height and the depth of snow, which make it impossible to approach the central ridge, except in a very few places.

The range consists of three parts: the Hindoo Coosh, or Indian Caucasus, which extends from the Parapamisan range in Afghanistan to Cashmere; the Himalaya, or Imaus of the ancients, which stretches from the valley of Cashmere to the sources of the Brahmapootra; and, lastly, the mountains of Bhotan and Assam,—the three making one magnificent unbroken chain.

The Hindoo Coosh, which has its name from a mountain of great height north of the city of Cabul, is very broad to the west, extending over many degrees of latitude, and, together with the offsets of the Beloot Tagh, fills the countries of Kafferistan, Koondez, and Budaksha. From the plains to the south it seems to consist of four distinct ranges running one above another, the last of which abuts on the table-land, and is so high that its snowy summits are visible at the distance of 150 miles. One of the ridges runs along the table-land parallel to the principal chain at the distance of 200 miles, known as the Ice Mountains, or Kara-Korum of the natives. Another ridge of stupendous height encloses the beautiful valley of Cashmere, to the east of which the chain takes the name of Himalaya, "the dwelling of snow," and extends 300 miles to the sources of the Brahmapootra, varying in breadth from 250 to 350 miles, and occupying an area of 600,000 square miles.

The general structure of the Himalaya is very regular; the first range of hills that rise above the plains of Hindostan is alluvial, north of which lies the Tariyani, a tract from 10 to 30 miles wide, 1000 feet above the sea, covered with dense, pestilential jungle, and extending along the foot of the range. North of this region are rocky ridges, 5000 or 6000 feet high. Between these and the higher ranges lie the peaceful and well-cultivated valleys of Nepaul, Bhotan, and Assam, of inexhaustible fertility, interspersed with picturesque and populous towns and villages. Though separated by mountain-groups, they form the principal terrace of the Himalaya, between the Sutlej and the Brahmapootra.

Behind these are mountains from 10,000 to 12,000 feet high, flanked by magnificent forests, and, lastly, the snowy ranges rise in succession to the table-land.

The principal and most elevated chains are cut by narrow, gloomy ravines and transverse dusky gorges, through which the torrents of melted snow rush to swell the rivers of Hindostan. The character of the valleys becomes softer in the lower regions, till at last the luxuriance of vegetation and beauty cannot be surpassed. Transverse valleys, however, are more frequent in the Hindoo Coosh than in the Himalaya, where they consist chiefly of such chasms filled with wreck as the tributaries of the Indus and Ganges have made in bursting through the chain.

The mean height of the Himalaya is stupendous, certainly not less than from 16,000 to 20,000 feet, though the peaks exceeding that elevation are not to be numbered, especially at the sources of the Sutlej; indeed, from that river to the Kalee the chain exhibits an endless succession of the loftiest mountains on earth: forty of them surpass the height of Chimborazo, the highest but one of the Andes, and many reach the height of 25,000 feet at least. So rugged is this part of the magnificent chain, that the military parade at Sabathoo, half a mile long, and a quarter of a mile broad, is said to be the only level ground between it and the Tartar frontier on the north, or the valley of Nepaul to the east. Towards the fruitful valleys of Nepaul and Bhotan the Himalaya is equally lofty, some of the mountains being from 25,000 to 28,000 feet high, but it is narrower, and the descent to the plains excessively rapid, especially in the territory of Bhotan, where the dip from the table-land is more than 10,000 feet in ten miles. The valleys are crevices so deep and narrow, and the mountains that hang over them in menacing cliffs are so lofty, that these abysses are shrouded in perpetual gloom, except when the rays of a vertical sun penetrate their depths. From the steepness of the descent the rivers shoot down with the swiftness of an arrow, filling the caverns with foam and the air with mist. At the very base of this wild region lies the elevated and peaceful valley of Bhotan, vividly green and shaded by magnificent forests. Another rapid descent of 1000 feet leads to the plain of the Ganges.

The Himalaya still maintains great height along the north of Assam, and at the sources of the Brahmapootra the parent stem and its branches extend in breadth over two degrees of latitude, forming a vast mountain knot, with summits 20,000 feet high. Beyond this point nothing certain is known of the range, but it or some of its branches are supposed to cross the southern provinces of the Chinese empire, and to end in the volcanic island of Formosa. Little more is known of the northern side of the mountains than that the passes are about 5000 feet above the plains of Tibet.

The passes over the Hindoo Coosh, though not the highest, are very formidable; there are six from Cabul to the plains of Turkistan, and so deep and so much enclosed are the defiles, that Sir Alexander Burnes never could obtain an observation of the pole star in the whole journey from Barmeean till within 30 miles of Turkistan.

Most of the passes over the Himalaya are but little lower than the top of Mont Blanc; many are higher, especially near the Sutlej, where they are from 18,000 to 19,000 feet high, and that north-east of Khoonawur is 20,000 feet above the level of the sea, the highest that has been attempted. All are terrific, and the fatigue and suffering from the rarity of the air in the last 500 feet is not to be described. Animals are as much distressed as human beings, and many die. Thousands of birds perish from the violence of the wind, the drifting snow is often fatal to travellers, and violent thunder-storms add to the horror of the journey. The Niti Pass, by which Mr. Moorcroft ascended to the sacred lake of Manasa in Tibet, is tremendous; he and his guide had not only to walk barefooted from the risk of slipping, but they were obliged to creep along the most frightful chasms, holding by twigs and tufts of grass, and sometimes they crossed deep and awful crevices on a branch of a tree, or loose stones thrown across; yet these are the thoroughfares for commerce in the Himalaya, never repaired nor susceptible of improvement from the frequent landslips and torrents.

The loftiest peaks being bare of snow gives great variety of colour and beauty to the scenery, which in these passes is at all times magnificent. During the day the stupendous size of the mountains, their interminable extent, the variety and sharpness of their forms, and, above all, the tender clearness of their distant outline melting into the pale blue sky, contrasted with the deep azure above, is described as a scene of wild and wonderful beauty. At midnight, when myriads of stars sparkle in the black sky, and the pure blue of the mountains looks deeper still below the pale white gleam of the earth and snow-light, the effect is of unparalleled solemnity, and no language can describe the splendour of the sunbeams at daybreak streaming between the high peaks, and throwing their gigantic shadows on the mountains below. There, far above the habitation of man, no living thing exists; no sound is heard; the very echo of the traveller's footsteps startles him in the awful solitude and silence that reigns in these august dwellings of everlasting snow.

Nature has in mercy mitigated the intense rigour of the cold in these high lands in a degree unexampled in other mountainous regions. The climate is mild, the valleys are verdant and inhabited, corn and fruit ripen at elevations which in other countries, even under the equator, would be buried in permanent snow.

It is also a peculiarity in these mountains, that the higher the range the higher likewise is the limit of snow and vegetation. On the southern slopes of the first range Mr. Gerard found cultivation 10,000 feet above the sea; in the valleys of the second range he met with shepherds feeding their flocks and dwelling at the height of 14,000 feet; and on the table-land of Tibet, the highest habitation of man in the Old World, the ground is cultivated at the altitude of 13,600 feet, which is only 2130 feet lower than the summit of Mont Blanc. In Chinese Tartary good crops of wheat are raised 16,000 feet above the sea; the vine and other fruits thrive in the valleys of these high plains. The temperature of the earth probably has some influence on the vegetation: as many hot springs exist in the Himalaya at great heights, there must be a source of heat below these mountains which in some places comes near the surface, and possibly may be connected with the volcanic fires in the central chains of the table-land. Hot springs abound in the valley of Jumnotra; and as it is well known that many plants thrive in very cold air if their roots are well protected, it may be the cause of pine-trees flourishing in that valley nearly 13,000 feet above the sea, and of the splendid forests of the deodar, a pine that grows to gigantic size even in the snow.

According to Captain and Mr. Gerard the line of perpetual congelation is at an elevation of only 12,800 feet on the southern slopes of the Himalaya, while on the northern side it is 15,600 feet above the sea—a remarkable circumstance, which is ascribed to the fogs that rise from the plains of Hindostan on one hand, and the serenity that prevails on the other: something may be due to radiation from the high northern plains, which, being so near, have much greater effect on the temperature than the warmer but more distant plains on the south.

Four vast secondary chains leave the Himalaya at the great mountain-knot at the sources of the Brahmapootra, in the Chinese province of Yunnan, and extend through the Indo-Chinese peninsula and the countries east of the Ganges, in a southern but diverging direction, leaving large and fertile kingdoms between them. The Birmano-Siamese chain is the most extensive, reaching to the extremity of the Malayan peninsula at Cape Romania, the most southerly point of Asia; it may be traced through the island of Sumatra parallel to the coast, and also in the islands of Banka and Beliton, where it ends.

Another range, called the Laos-Siamese chain, forms the eastern boundary of the kingdom of Siam, and the Annamatic chain, from the same origin, separates the empire of Annam from Tonquin and Cochin China.

These slightly diverging lines of mountains yield gold, silver, tin, of the best quality, in great plenty, almost on the surface, and precious stones, as rubies and sapphires. Mountains in low latitudes have nothing of the severe character of those in less favoured climes. Magnificent forests reach their summit; spices, dyes of brilliant tints, medicinal and odoriferous plants clothe these declivities; and in the low grounds the fruits of India and China grow in perfection in a soil which yields three crops of grain in the year.

The crest of the Himalaya is of stratified crystalline rocks, especially gneiss, with large granitic veins, and beds of quartz of huge magnitude. The zone, between 15,000 and 18,000 feet above the level of the sea, is of silurian strata, below which sandstone prevails: granite is most frequent at the base, and probably forms the foundation of the chain. Strata of comparatively modern date occur at great elevations. These sedimentary formations, prevailing also on the acclivities of the Alps and Apennines, show that the epochs of elevation in parts of the earth widely remote from one another, if not simultaneous, were at least not very different. There can be no doubt that very great geological changes have taken place at a comparatively recent period in the Himalaya, and through an extensive part of the Asiatic continent.

The Altaï mountains, which form the northern margin of the table-land, are unconnected with the Ural chain: they are separated from it by 400 miles of a low marshy country, part of the steppe of the Kirghiz, and by the Dalai mountains, a low range never above 2000 feet high, which runs between the 64th meridian and the left bank of the Irtysh. The Altaï chain begins on the right bank of that river at the north-west angle of the table-land, and extends in a serpentine line to the Pacific, south of the Gulf of Okhotzk, dividing the high lands of Tartary and China from the wastes of Asiatic Siberia. Under the name of the Aldan Mountains it skirts the north-west side of the Gulf of Okhotzk, and then stretches to Behring's Straits, its length being 4500 miles. The breadth of this chain varies from 400 to 1000 miles, but towards the 105th meridian it is contracted to about 150, by a projection of the desert of the Great Gobi. Its height bears no proportion to its length and breadth. Indeed the Little Altaï, the only part of the chain properly so called lying between the Irtysh and the 86th degree of east longitude, can only be regarded as a succession of terraces of a swelling outline, descending by steps from the table-land, and ending in promontories on the Siberian plains. There are numerous large lakes on these terraces and on the mountain valleys, as in the mountain systems of Europe. The general form of this part of the chain is monotonous from the prevalence of straight lines and smooth rounded outlines. Long ridges with flattened summits, or small table-lands, not more than 6000 feet high, is their usual structure, rarely attaining the line of perennial congelation: snow however is permanent on the Korgon table-land, 9900 feet above the sea, supposed to be the culminating point of this part of the chain. These table-lands bear a strong resemblance to those in the Scandinavian mountains in baldness and sterility, but their flanks are clothed with forests, verdant meadows, and pastoral valleys.

East of the 86th meridian this region of low mountains splits into three branches, enclosing longitudinal valleys for 450 miles. The central chain, called the Tongnou Oola, may be regarded as the principal continuation of the Altaï: it lies nearly along the

50th parallel of latitude, but, bending northwards, passes between the lakes Kossagol and Baikal, under the name of the Sayansk Mountains. The granite range of the Baikal, properly so called, meets the Sayansk chain nearly at right angles, and unites it with the mountains of the Upper Angara. At the point where the axes of the Baikal and Sayansk chains cross, the mountains are highest, and there only the Altaï assumes the form of a regular chain. The principal part of the Baikal group is 500 miles long, from 10 to 60 wide, high and snow-capped, but without glaciers. It flanks Lake Baikal on the north, the largest of Alpine lakes, so imbedded in a knot of mountains, partly granitic, partly volcanic, that rocks and pillars of granite rise from its bed. The mountains south of the lake are but the face of the table-land; a traveller ascending them finds himself at once in the desert of Gobi, which stretches in unbroken sadness to the Great Wall of China.

The Daouria Mountains, a volcanic portion of the Altaï, which borders the table-land on the north-east, follow the Baikal chain; and farther east, at the sources of the Aldan, the Altaï range takes the name of the Yablonnoi Khrebet, and stretches south of the Gulf of Okhotzk to the coast of the Pacific, opposite to the island of Tarakai; while another part, 1000 miles broad, fills the space between the Gulf of Okhotzk and the river Lena, and then, bending to the north-east, ends in the peninsula of Kamtschatka.

A great portion of the Altaï chain is unknown to Europeans; the innumerable branches that penetrate the Chinese empire are completely so: those belonging to Russia abound in a great variety of precious and rare metals and minerals—silver, copper, and iron. In the Yablonnoi range and other parts there are whole mountains of porphyry, with red and green jasper; coal is also found; and in a branch of the Altaï, between the rivers Obi and Yenissei, there are mines of coal which were set on fire by lightning, and have continued to burn more than a century. The Siberian mountains far surpass the Andes in the richness of their gold-mines. The eastern flank of the Ural chain, and some of the northern spurs of the Altaï, have furnished an immense quantity, but a region as large as France has lately been discovered in Siberia covered with the richest gold alluvium, lying above rocks filled with that precious metal. The mines of the Ural and Altaï are in metamorphic schists adjacent to the greenstones, syenites, and serpentines that have caused their change; and as the same formation prevails throughout the greater part of the Altaï and Aldan chains almost to Kamtschatka, there is every reason to believe that the whole of that vast region is auriferous: besides, as many of the northern offsets of the Altaï are particularly rich, it may be concluded that the southern branches in the Chinese empire are equally so. Thus all southern Siberia and Chinese Tartary form an auriferous district probably greater than all Europe, which extends even to our dominions in Hindostan, where the gold formations are unexplored.

The sedimentary deposits in this extensive mountain range are more ancient than the granite, syenite, and porphyries; consequently these igneous rocks have not here formed part of the original crust of the globe. Rocks of the Paleozoic series occupy the greater part of the Altaï, and probably there are none more modern. There are no volcanic rocks, ancient or modern, west of the Yenesei, but they abound to the east of that river, even to Kamtschatka, which is full of them.

The physical characters and the fossil remains of this extensive mountain system have little relation with the geological formations of Europe and America. Eastern Siberia seems even to form an insulated district by itself, and that part between the town of Yakoutzk and the mouth of the Lena appears to have been raised at a later period than the part of Siberia stretching westward to the Sayanok Mountains: moreover the elevation of the Little Altaï was probably contemporaneous with that of the Ural Mountains.

Little more is known of the eastern boundary of the table-land of Tibet than that between the sources of the Brahmapootra and the Altaï chain nearly a million of square miles of the Chinese empire are covered with mountains, which begin under the 98th meridian at the edge of the table-land, and descend to the 112th degree of east longitude in southern China, and to the 114th degree in the north. The eastern boundary of this mountainous region is said to be the chains of the In-Shan and Khing-Khan Oolas. The former begins at the southern extremity of Tartary, near the Yellow River, and maintains a very tortuous course to the snow-clad mountains of Petsha, 15,000 feet high. It then goes north, under the name of the Khing-Khan Oola, in a serrated granitic chain, separating the table-land of Mongolia from the country of the Manchoux, and joins the Yablonnoi branch of the Altaï at right angles about the 55th degree of north latitude.

The table-land of Tibet is only 4000 feet above the sea towards the north, but it rises in Little Tibet to between 11,000 and 12,000 feet. The Kuen-luen, the most southerly of the two diagonal mountain-chains that cross the table-land, begins at the Hindoo Coosh, in latitude 35° 30′, and extends eastward in two branches, which again unite in the K'han of eastern Tibet, nearly in the centre of the table-land, where they form an elevated mountain plain round the Lake of Koko-Nor, from whence those immense mountain-ranges diverge which render the south-western provinces of China the most elevated region on earth. The country of Tibet lying between the Himalaya and the Kuen-luen consists of rocky mountainous ridges, extending from N.W. to S.E., separated by long valleys, in which flow the upper courses of the Brahmapootra, Sutlej, and Indus. According to Mr. Moorcroft, the sacred lake Manasa, in Great Tibet, and the surrounding country, is 17,000 feet above the sea, which is 1270 feet higher than Mont Blanc. In this elevated region the sheltered valleys and the borders of the streams alone are available for agriculture; and as the summer sun is powerful, wheat and barley grow, and many of the fruits of southern Europe ripen. The city of H'Lassa, in eastern Tibet, the residence of the Grand Lama, is surrounded by vineyards, and is called by the Chinese the "Realm of Pleasure." There are no trees in this country, and the ground in cultivation bears a small proportion to the grassy steppes, which extend in endless monotony, grazed by thousands of the shawl-wool goats, sheep, and cattle. There are many lakes in the table-land; some in Ladok contain borax, a salt very useful in the arts, found only here and at Corbali in Tuscany, and the Lipari islands.

In summer the sun is powerful at midday, the air is of the purest transparency, and the azure of the sky so deep that it seems black as in the darkest night. The rising moon does not enlighten the atmosphere, no warning radiance announces her approach, till her limb touches the horizon, and the stars shine with the distinctness and brilliancy of suns. In southern Tibet the verdure is confined to favoured spots, the bleak mountains and high plains are sternly gloomy—a scene of barrenness not to be conceived. Solitude reigns in these dreary wastes, where there is not a tree nor even a shrub to be seen of more than a few inches high. The scanty short-lived verdure vanishes in October, the country looks as if fire had passed over it, and cutting dry winds blow with irresistible fury, howling in the bare mountains, whirling the snow through the air, and freezing to death the unfortunate traveller benighted in their defiles.

Yarkand and Khotan, provinces of Chinese Tartary, which lie beyond the two diagonal chains, are less elevated and more fertile than Tibet. They are watered by five rivers, and contain several large cities; Yarkand, the most considerable of these, is the emporium of commerce between Tibet, Turkistan, China, and Russia. Gold, rubies, silk, and other productions are exported.

The Tartar range of the Thian-Shan is very high; the Bogda Oola, or Holy Mountain, near Lake Lop, its highest point, is always covered with snow; and it has two active volcanoes, one on each side—a solitary instance of volcanic vents so far from the sea. This range runs along the 42nd parallel of north latitude, forming at its western extremity a mountain-knot with the Beloot Tagh, in the centre of which lies the small table-land of Pamere, 15,600 feet high, called by the natives the "Roof of the World." Its remarkable elevation was first observed by the enterprising Venetian traveller, Marco Paolo, six centuries ago. The Oxus originates in a glacier of the Pooshtee Khur, a peak of the Beloot Tagh, near the plain of Pamere; and the lake Sir-i-Kol is here the source of the Yarkand, and the Kokan also rises from this plain, which is intensely cold in winter, and in summer is alive with flocks of sheep and goats.

Zungary, or Mongolia, the country between the Thian-Shan and the Altaï, is hardly known further than that its grassy steppes, intersected by many lakes and offsets from the Altaï, are the pasture-grounds of the wandering Kirghis.

The remarkable feature of the table-land is the desert of the Great Gobi, which occupies an area of 300,000 square miles in its eastern extremity, interrupted only by a few spots of pasture and low bushes. Wide tracts are flat and covered with small stones or sand, and at a great distance from one another there are low hills, destitute of wood and water; its general elevation is about 4000 feet above the sea, but it is intersected from west to east by a depressed valley aptly named Shamo, or the "Sea of Sand," which is also mixed with salt. West from it lies the Han-Hai, the "Dry Sea," a barren plain of shifting sand blown into high ridges. Here, as in all deserts, the summer sun is scorching, the winter's cold intolerable. All the plains of Mongolia are intensely cold, because the hills to the north are too low to screen them from the polar blast, and, being higher than the Siberian deserts, they are bitterly cold; no month in the year is free from frost and snow, yet it is not deep enough to prevent cattle from finding pasture. Sandy deserts like that of the Great Gobi occupy much of the country south of the Chinese branches of the Altaï.

Such is the stupendous zone of high land that girds the old continent throughout its whole length. In the extensive plains on each side of it several independent mountain systems rise, though much inferior to it in extent and height.

CHAPTER V.

SECONDARY MOUNTAIN SYSTEMS OF THE GREAT CONTINENT—THAT OF SCANDINAVIA—GREAT BRITAIN AND IRELAND—THE URAL MOUNTAINS—THE GREAT NORTHERN PLAIN.

THE great northern plain is broken by two masses of high land, in every respect inferior to those described: they are the Scandinavian system and the Ural Mountains, the arbitrary limit between Europe and Asia.

The range of primary mountains which has given its form to the Scandinavian peninsula begins at Cape Lindesnaes, the most southerly point of Norway, and, after running along its western coast 1000 miles in a north-easterly direction, ends at Cape Nord Kyn on the Polar Ocean, the extremity of Europe. The highest elevation of this chain is not more than 8412 feet. It has been compared to a great wave or billow, rising gradually from the east, which, after having formed a crest, falls perpendicularly into the sea in the west. There are 3696 square miles of this peninsula above the line of perpetual snow.

The southern portion of the chain consists of ridges following the general direction of the range, 150 miles broad. At the distance of 360 miles from Cape Lindesnaes the mountains form a single elevated mass, terminated by a table-land, which maintains an altitude of 4500 feet for 100 miles. It slopes towards the east, but plunges at once in high precipices into a deep sea on the west.

The surface is barren, marshy, and bristled with peaks; besides, an area of 600 square leagues is occupied by the Snae Braen, the greatest mass of perpetual snow and glaciers on the continent of Europe. A prominent cluster of mountains follows, from whence a single chain, 25 miles broad, maintains an uninterrupted line to the island of Megaree, where it terminates in North Cape, a huge barren rock perpetually lashed by the surge of the Polar Ocean. Offsets from these mountains cover Finland and the low rocky table-land of Lapland: the valleys and countries along the eastern side of the chain abound in forests and Alpine lakes.

The iron-bound coast of Norway is a continued series of rocky islands, capes, promontories, and precipitous cliffs, rent into chasms which penetrate miles into the heart of the mountains. These chasms, or fiords, are either partly or entirely filled by arms of the sea; in the former case the shores are fertile and inhabited. Fiords are not peculiar to the coast of Norway: they are even more extensive in Greenland and Iceland, and of a more stern character, overhung by snow-clad rocks and glaciers.

As the Scandinavian mountains, those of Feroe, Britain, Ireland, and the north-eastern parts of Iceland have a similar character, and follow the same general directions, they must have been elevated by forces acting in parallel lines, and therefore may be regarded as belonging to the same system.

The Feroe islands, due west from Norway, rise at once in a table-land 2000 feet high, bounded by precipitous cliffs, which dip into the ocean. Some parts of these islands are gradually sinking below their former level; indeed there seems to be an extraordinary flexibility in the crust of the earth in these high northern latitudes: it is bending below its former level in south Sweden, Feroe, and the west coast of Greenland, or in a zone between the 55th and 62nd or 63rd parallels, while the coast of Norway is rising at the rate of four feet in a hundred years from Sölvitsberg northward to Lapland, where the elevation is greatest.

The rocky islands of Zetland and those of Orkney form part of the mountain system of Scotland: the Orkney islands have evidently been separated from the mainland by the Pentland Firth, where the currents run with prodigious violence. The north-western part of Scotland is a table-land from 1000 to 2000 feet high, which ends abruptly in the sea, covered with heath, peat-mosses, and pasture. The general direction of the Scottish mountains, like those of Scandinavia, is from north-east to south-west, divided by a long line of lakes in the same direction, extending from the Moray Firth completely across the island to south of the island of Mull. Lakes of the most picturesque beauty abound among the Scottish mountains. The Grampian hills, with their offsets and some low ranges, fill the greater part of Scotland north of the Clyde and Forth. Ben Nevis, only 4374 feet above the sea, is the highest hill in the British islands.

The east coast of Scotland is generally bleak, though in many parts it is extremely fertile, and may be cited as a model of good cultivation; and the midland and southern counties are not inferior either in the quality of the soil or the excellence of the husbandry. To the west the country is wildly picturesque; the coast of the Atlantic, penetrated by the sea, which is covered with islands, bears a strong resemblance to that of Norway.

There cannot be a doubt that the Hebrides formed part of the mainland at some remote geological period, since they follow the direction of the mountain system in two parallel lines of rugged and imposing aspect, never exceeding the height of 3200 feet. The undulating country on the borders of Scotland becomes higher in the west of England and North Wales, where the hills are wild, but the valleys are cultivated like a garden, and the English lake scenery is of the most gentle beauty.

Evergreen Ireland is mostly a mountainous country, and opposes to the Atlantic storms an iron-bound coast of the wildest aspect; but it is rich in arable land and pasture, and possesses the most picturesque lake-scenery; indeed, fresh-water lakes in the mountain valleys, so peculiarly characteristic of the European system, are the great ornaments of the high lands of Britain.

Various parts of the British islands were dry

land while most of the continent of Europe was yet below the ancient ocean. The high land of Lammermuir, the Grampian hills in Scotland, and those of Cumberland in England, were raised before the Alps had begun to appear above the waves. In general all the highest parts of the British mountains are of granite and stratified crystalline rocks. The primary fossiliferous strata are of immense thickness in Cumberland and in the north of Wales, and the old red sandstone, many hundred feet thick, stretches from sea to sea along the flanks of the Grampians. The coal-strata are developed on a great scale in the south of Scotland and the north of England, and examples of every formation, with one exception, are to be found in these islands. Volcanic fires had been very active in early times, and nowhere is the columnar structure more beautifully exhibited than in Fingal's Cave and the Storr of Sky in the Hebrides; and in the north of Ireland a base of 800 square miles of mica slate is covered with volcanic rocks, which end on the coast in the magnificent columns of the Giant's Causeway.

The Ural chain, the boundary between Europe and Asia, is the only interruption to the level of the great northern plain, and is altogether unconnected with, and far separated from, the Altaï Mountains by salt lakes, marshes, and deserts. The central ridge may be traced from between the Lake of Aral and the Caspian Sea; but as a chain it really begins on the right bank of the Ural river at the steppes of the

Kirghis, about the 51st degree of north latitude, and runs due north in a long narrow ridge to the Gulf of Kara in the Polar Ocean, though it may be said to terminate in dreary rocks on the west side of Nova Zembla. The Ural range is about the height of the mountains in the Black Forest or the Vosges, and, with few exceptions, is wooded to the top, chiefly by the pinus cimbra. The immense mineral riches of these mountains—gold, platina, magnetic iron, and copper—lie on the Siberian side, and chiefly between the 54th and 60th degrees of north latitude, the only part that is colonized, and one of the most industrious and civilized regions of the Russian empire. To the south the chain is pastoral, about 100 miles broad, consisting of longitudinal ridges, the highest of which does not exceed 3498 feet; in this part diamonds are found. To the north of the mining district the narrow mural mass, which is at most but 5720 feet above the sea-level, is covered with impenetrable forests and deep morasses, altogether uninhabitable and unexplored. Throughout the Ural Mountains there are neither precipices, transverse gorges, nor any of the characteristics of a high chain: the descent on both sides is so gentle that in many places it is difficult to know where the plain begins; and the road over the chain from Russia to Siberia by Ekaterinburg is so low that it hardly seems to be a mountain pass. The gentle descent and sluggishness of the streams produce extensive marshes along the Siberian base of the range. To the arduous and enterprising researches of Sir

Roderick Murchison we are indebted for almost all we know of these mountains; he found them on the western side to be composed of silurian, devonian, and carboniferous rocks more or less altered and crystallized; and on the eastern side the mines are in metamorphic strata, mixed with rocks of igneous origin, and the central axis is of quartzose and chloritic rocks.

The great zone of high land which extends along the old continent from the Atlantic to the shores of the Pacific Ocean divides the low lands into two very unequal parts. That to the north, only broken by the Ural range, and the Valdai table-land of still less elevation, stretches from the Thames or the British hills and the eastern bank of the Seine to Behring's Straits, including more than 190° of longitude, and occupying an area of at least four millions and a half of square geographical miles, which is a third more than all Europe. The greater part of it is perfectly level, with a few elevations and low hills, and in many places a dead level extends hundreds of miles. The country between the Carpathian and Ural Mountains is a flat, on which there is scarcely a rise in 1500 miles, and in the steppes of southern Russia and Siberia the extent of level ground is immense. The mean absolute height of the flat provinces of France is 480 feet; Moscow, the highest point of the European plain, is also 480 feet high, from whence the land slopes imperceptibly to the sea both on the north and south, till it absolutely dips below its level. Holland, on one side, would be overflowed were it not for its dykes, and towards Astrakan the plain sinks still lower. The whole of that extensive country north and east of the Caspian Sea, and around the lake of Aral, forms a vast cavity of 18,000 square leagues, all considerably below the level of the ocean; and the surface of the Caspian Sea itself, the lowest point, has a depression of 348 feet.

The European part of the plain is highly cultivated and very productive in the more civilized countries in its western and middle regions and along the Baltic. The greatest amount of cultivated land lies to the north of the watershed which stretches from the Carpathians to the centre of the Ural chain; yet there are large heaths which extend from the extremity of Jutland through Lunebourg and Westphalia to Belgium. The land is of excellent quality to the south of it. Round Polkova and Moscow there is an extent of the finest vegetable mould, equal in size to France and the Spanish peninsula together, which forms part of the High Steppe, and is mostly in a state of nature.

A large portion of the great plain is pasture-land, and wide tracts are covered with natural forests, especially in Poland and Russia, where there are millions of acres of pine, fir, and deciduous trees.

The quantity of waste land in Europe is very great, and there are also many swamps; a morass as long as England extends along the 52nd parallel of latitude, following the course of the river Prepit, a branch of the Dniestre, which runs through its centre. There are swamps at the mouths of many of the sluggish rivers in central Europe; they cover 1970 square miles in Denmark, and mossy quagmires occur frequently in the more northerly parts.

Towards the eastern extremity of Europe the great plain assumes the peculiar character of desert called a steppe, a word supposed to be of Tartar origin, signifying a level waste destitute of trees; hence the steppes may vary according to the nature of the soil. They begin at the river Dnieper, and extend along the shores of the Black Sea: they include all the country north and east of the Caspian Lake and Independent Tartary, and, passing between the Ural and Altaï Mountains, they may be said to occupy all the low lands of Siberia. Hundreds of leagues may be traversed east from the Dnieper without variation of scene; a dead level of thin but luxuriant pasture, bounded only by the horizon, day after day the same unbroken monotony fatigues the eye: sometimes there is the appearance of a lake, which vanishes on approach, the phantom of atmospheric refraction. Horses and cattle beyond number give some animation to the scene so long as the steppes are green, but winter comes in October, and then they become a trackless field of spotless snow. Fearful storms rage, and the dry snow is driven by the gale with a violence which neither man nor animal can resist, while the sky is clear and the sun shines cold and bright above the earthly turmoil.

The contest between spring and winter is long and severe, for—

> "Winter oft at once resumes the breeze,
> Chills the pale morn, and bids his driving sleets
> Deform the day, delightless."

Yet when gentler gales succeed, and the waters run off in torrents through the channels which they cut in the soft ground, the earth is again verdant. The scorching summer's sun is as severe in its consequences in these wild regions as the winter's cold: in June the steppes are parched, no shower falls, nor does a drop of dew refresh the thirsty and rent earth: the sun rises and sets like a globe of fire, and during the day he is obscured by a thick mist from the evaporation. In some seasons the drought is excessive; the air is filled with dust in impalpable powder; the springs become dry, and cattle perish in thousands. Death triumphs over animal and vegetable nature, and desolation tracks the scene to the utmost verge of the horizon, a hideous wreck.

Much of this country is covered by an excellent but thin soil, fit for corn, which grows luxuriantly wherever it has been tried; but a stiff cold clay at a small distance below the surface kills every herb that has deep roots, and no plants thrive but those which can resist the extreme vicissitudes of climate. A very wide range is hopelessly barren; the country from the Caucasus along the shores of the Black and Caspian Seas, a dead flat twice the size of the British islands, is desert and destitute of fresh water. An efflorescence of salt covers the surface like hoar-frost;

even the atmosphere and the dew are saline, and many salt-lakes in the neighbourhood of Astrakan furnish great quantities of common salt and nitre. Saline plants, with patches of verdure few and far between, are the only signs of vegetable life, but about Astrakan there is soil and cultivation. Some low hills occur in the country between the Caspian and the Lake of Aral, but it is mostly an ocean of shifting sand, often driven by appalling whirlwinds.

Turkistan is a sandy desert, except on the banks of the Oxus and the Jaxartes, and as far on each side of them as canals convey the fertilizing waters. To the north barrenness gives place to verdure between the Ural river and the terraces and mountains of central Asia, where the steppes of the Kirghiz afford pasture to thousands of camels and cattle belonging to these wandering hordes.

Siberia is either a dead level or undulating surface of more than 7,000,000 of square miles, between the North Pacific and the Ural Mountains, the Polar Sea and the Altaï range, whose terraces and offsets end in those plains, like headlands and promontories in the ocean. M. Middendorf, indeed, met with a chain of most desolate mountains on the shores of the Polar Ocean, in the country of the Samoides; and the almost inapproachable coast far to the east is unexplored. The mineral riches of the mountains have brought together a population who inhabit towns of considerable importance along the base of the Ural and Altaï chains, where the ground yields good crops and pasture; and there are forests on the undulations of the mountains and on the plains. There are many hundred square miles of rich black mould covered with trees and grass, uninhabited, between the river Tobal and the upper course of the Obi, within the limit where corn would grow; but even this valuable soil is studded with small lakes of salt and fresh water, a chain of which, 300 miles long, skirts the base of the Ural Mountains.

North of the 62nd parallel of latitude corn does not ripen, on account of the biting blasts from the Icy Ocean which sweep supreme over these unprotected wastes. In a higher latitude even the interminable forests of gloomy fir are seen no more; all is a wide-spreading desolation of salt steppes, boundless swamps, and lakes of salt and fresh water. The cold is so intense there that the spongy soil is perpetually frozen to the depth of some hundred feet below the surface; and the surface itself, not thawed before the end of June, is again ice-bound by the middle of September, and deep snow covers the ground nine or ten months in the year. Happily gales of wind are not frequent during winter, but when they do occur no living thing ventures to face them. The sun, though long absent from these dismal regions, does not leave them to utter darkness; the extraordinary brilliancy of the stars, and the gleaming snow-light, produce a kind of twilight, which is augmented by the splendid coruscations of the Aurora Borealis.

The scorching heat of the summer's sun produces a change like magic on the southern provinces of the Siberian wilderness. The snow is scarcely gone before the ground is covered with verdure, and flowers of various hues blossom, bear their seed, and die in a few months, when winter resumes his empire. A still shorter-lived vegetation scantily covers the plains in the far north, and, on the shores of the Icy Ocean, even reindeer-moss grows scantily.

The abundance of fur-bearing animals in the less rigorous parts of the Siberian deserts has tempted the Russians to colonize and build towns on these frozen plains. Yakutsk, on the river Lena, in 62° 1′ 30″ N. latitude, is probably the coldest town on earth. The ground is perpetually frozen to the depth of more than 400 feet, of which three feet only are thawed in summer, when Fahrenheit's thermometer is frequently 77° in the shade; and as there is sometimes no frost for four months, larch forests cover the ground, and wheat and rye produce from fifteen to forty fold. In winter the cold is so intense that mercury is constantly frozen two months, and occasionally even three.

In the northern parts of Europe the silurian, devonian, and carboniferous strata are widely developed, and more to the south they are followed in ascending order by immense tracts of the higher series of secondary rocks, abounding in the huge monsters of a former world. Very large and interesting tertiary basins fill the ancient hollows in many parts of the plain, which are crowded with the remains of animals that no longer exist. Of these the most important are the London, Paris, Brussels, and Moscow basins, with many others in the north of Germany and Russia, and alluvial soil covers the greater part of the plain. In the east Sir Roderick Murchison has determined the boundary of a region twice as large as France, extending from the Polar Ocean to the southern steppes, and from beyond the Volga to the flanks of the Ural chain, which consists of a red deposit of sand and marl, full of copper in grains, belonging to the Permian system. This, and the immense tract of black loam already mentioned, are the principal features of eastern Europe.

CHAPTER VI.

THE SOUTHERN LOW LANDS OF THE GREAT CONTINENT, WITH THEIR SECONDARY TABLE-LANDS AND MOUNTAINS.

THE low lands to the south of the great mountain girdle of the old continent are much broken by its offsets, by separate groups of mountains, and still more by the deep indentation of bays and large seas. Situate in lower latitudes, and sheltered by mountains from the cutting Siberian winds, these plains are of a more tropical character than those to the north; but they are strikingly contrasted in their different parts,—either rich in all the exuberance that heat, moisture, and soil can produce, or covered by wastes of bare sand,—in the most advanced state of cultivation, or in the wildest garb of nature.

The barren parts of the low lands lying between the eastern shores of China and the Indus bear a small proportion to the riches of a soil vivified by tropical warmth, and watered by the periodical inundations of the mighty rivers that burst from the icy caverns of Tibet and the Himalaya. On the contrary, the favoured regions on that part of the low lands lying between the Persian Gulf, the Euphrates, and the Atlas Mountains, are small when compared with the immense expanse of the Arabian and African deserts, calcined and scorched by an equatorial sun. The blessing of a mountain zone, pouring out its everlasting treasures of moisture, the life-blood of the soil, is nowhere more strikingly exhibited than in the contrast formed by these two regions of the globe.

The Tartar country of Mandshur, watered by the river Amour, but little known to Europeans, lies immediately south of the Ўablonnoi branch of the Altaï chain, and consequently partakes of the desert aspect of Siberia, and, in its northern parts, even of the Great Gobi. It is partly intersected by mountains, and covered by dense forests; nevertheless, oats grow in the plains, and even wheat in sheltered places. Towards Corea the country is more fertile; in that peninsula there are cultivated plains at the base of its central mountain range.

China is the most productive country on the face of the earth; an alluvial plain of 210,000 square miles, formed by one of the most extensive river systems in the old world, occupies its eastern part. This plain, seven times the size of Lombardy, is no less fertile, and perfectly irrigated by canals. The great canal traverses the eastern part of the plain for 700 miles, of which 500 are in a straight line of considerable breadth, with a current in the greater part of it. Most part of the plain is in rice and garden ground, the whole cultivated with the spade. The tea-plant grows on a low range of hills between the 30th and 22nd parallels of north latitude, an offset from the Pe-ling chain. The cold in winter is much greater than in corresponding European latitudes, and the heat in summer is proportionally excessive.

The Indo-Chinese peninsula, lying between China and the river Brahmapootra, has an area of 77,700 square miles, and projects 1500 miles into the ocean. The plains lying between the offsets descending from the east end of the Himalaya, and which divide it longitudinally, as before mentioned, are very extensive. The Birman empire alone, which occupies the valley of the Irrawaddy, is said to be as large as France, and not less fertile, especially its southern part, which is the granary of the empire. Magnificent rivers intersect the alluvial plains, whose soil they have brought down from the table-land of Tibet, and still continue to deposit in great quantities in the deltas at their mouths.

The plains of Hindostan extend 2000 miles along the southern slope of the Himalaya and Hindoo Coosh, between the Brahmapootra and the Indus, and terminate on the south in the Bay of Bengal, the table-land of the Decan, and the Indian Ocean—a country embracing in its range every variety of climate, from tropical heat and moisture to the genial temperature of southern Europe.

The valley of the Ganges is one of the richest on the globe, and contains a greater extent of vegetable mould, and of land under cultivation, than any other country in this continent, except perhaps the Chinese empire. In its upper part, Sirhind and Delhi, the seat of the ancient Mongol empire, still rich in splendid specimens of Indian art, are partly arid, although in the latter there is fertile soil. The country is beautiful where the Jumna and other streams unite to form the Ganges. These rivers are often hemmed in by rocks and high banks, which in a great measure prevent the periodical overflow of the waters; this, however, is compensated by the coolness and moisture of the climate. The land gradually improves towards the east, as it becomes more flat, till at last there is not a stone to be seen for hundreds of miles down to the Gulf of Bengal. Wheat and other European grain is produced in the upper part of this magnificent valley, while in the south every variety of Indian fruit, rice, cotton, indigo, opium, and sugar, are the staple commodities. The ascent of the plain of the Ganges from the Bay of Bengal is so gradual, that Saharampore, nearly at the foot of the Himalaya, is only 1100 feet above the level of Calcutta; the consequence of which is, that the Ganges and Brahmapootra, with their branches, in the rainy season between June and September, lay Bengal under water for hundreds of miles in every direction, like a great sea. When the water subsides, the plains are verdant with rice and other grain; but when harvest is over, and the heat intense, the scene is changed—the country, divested of its beauty, becomes parched and dusty everywhere, except in the extensive jungles. It has been estimated that one-third of the British territory in India is covered with these rank marshy tracts.

The peninsula of Hindostan is occupied by the triangular-shaped table-land of the Decan, which is much lower, and totally unconnected with the table-land of Tibet. It has the primary ranges of the Ghauts on the east and west, and the Vendhya Mountains on the north, sloping by successive levels to the plains of Hindostan Proper. The surface of the Decan, between 3000 and 4000 feet above the sea, is a combination of plains, ridges of rock, and insulated flat-topped hills, which are numerous, especially in its north-eastern parts. These solitary and almost inaccessible heights rise abruptly from the plains, with all but perpendicular sides, which can only be scaled by steps cut in the rock, or by very dangerous paths. Many are fortified, and were the strongholds of the natives, but they never have withstood the determined intrepidity of British soldiers.

The peninsula terminates with the table-land of the Mysore, 7000 feet above the sea, surrounded by hills 1500 higher.

The base of this plateau, and indeed of all the Decan, is granite, and there are also syenitic and trap rocks, with abundance of primary and secondary fossiliferous strata. Though possessing the diamond-mines of Golconda, the true riches of this country consist in its vegetable mould, which in the Mysore is a hundred feet thick, an inexhaustible source of fertility. The sea-coasts on the two sides of the peninsula are essentially different: that of Malabar is rocky, but in many parts well cultivated, and its high mountains are covered with forests; whereas on the Coromandel coast the mountains are bare, and the wide maritime plains are for the most part parched.

The island of Ceylon, nearly equal in extent to Ireland, is almost joined to the southern extremity of the peninsula by sandbanks and small islands, between which the water is only six feet deep in spring tides. The Sanscrit name of the "Resplendent" may convey some idea of this island, rich and fertile in soil, adorned by lofty mountains, numerous streams, and primeval forests; in addition to which it is rich in precious stones, and has the pearl-oyster on its coast.

The Asiatic low lands are continued westward from the Indian peninsula by the Punjab and the Great Indian Desert. The Punjab, or "country of the five rivers," lies at the base of the Hindoo Coosh. Its most northern part consists of fertile terraces, highly cultivated, and valleys at the foot of the mountains. It is very productive in the plain within the limits of the periodical inundations of the rivers, and where it is watered by canals; in other parts it is pastoral. Lahore occupies the chief part of the Punjab; and the city of that name on the Indus, once the rival of Delhi, lies on the high road from Persia to India, and was made the capital of the kingdom by Runjeet Sing. The valley of the Indus throughout partakes of the character of the Punjab; it is fertile only where it is within reach of water; much of it is delta, which is occupied by rice-grounds; the rest is pasture, or sterile sal marshes.

South of the Punjab, and between the fertile plains of Hindostan and the left banks of the Indus, lies the Great Indian Desert, which is about 400 miles broad, and becomes more and more arid as it approaches the river. It consists of a hard clay, covered with shifting sand, driven into high waves by the wind, with some parts that are verdant after the rains. In the province of Cutch, south of the desert, a space of 7000 square miles, known as the Run of Cutch, is alternately a sandy salt desert and an inland sea. In April the waves of the sea are driven over it by the prevailing winds, leaving only a few grassy eminences, the resort of wild asses. The Desert of Mekram, an equally barren tract, extends along the Gulf of Oman from the mouths of the Indus to the Persian Gulf; in some places, however, it produces the Indian palm and the aromatic shrubs of Arabia Felix.

The scathed shores of the Arabian Gulf, where not a blade of grass freshens the arid sands, and the not less barren valley of the Euphrates and Tigris, except where the floods of these rivers irrigate the soil, separate Asia from Arabia and Africa, the most desert regions in the old world.

The peninsula of Arabia, divided into two parts by the tropic of Cancer, is about four times the size of France. No rivers, and few streams or springs, nourish this thirsty land, whose barren sands are scorched by a fierce sun. The central part is a tableland of moderate height, which, however, is said to have an elevation of 8000 feet in the province of

Haudramaut. To the south of the tropic it is an almost interminable ocean of drifting sand, wafted in clouds by the gale, and dreaded even by the wandering Beduin. At wide intervals, long, narrow depressions cheer the eye with brushwood and verdure. More to the north, mountains and hills cross the peninsula from S.W. to N.E., enclosing cultivated and fine pastoral valleys, adorned by groves of the date-palm and aromatic shrubs. Desolation once more resumes its domain where the table-land sinks into the Syrian desert, and throughout the rest of its circumference it descends in terraces or parallel ranges of mountains and hills to a flat sandy coast, from 30 to 100 miles wide, which surrounds the greater part of the peninsula, from the mouths of the Euphrates to the isthmus of Suez. The hills come close to the beach in the province of Oman, which is traversed by chains, and broken into piles of arid mountains, not more than 3500 feet high, with the exception of the Jebel Okkdar, which is 6000 feet above the sea, and is cleft by temporary streams and fertile valleys. Here the ground is cultivated and covered with verdure, and still farther south there is a line of oases fed by subterraneous springs, where the fruits common to Persia, India, and Arabia are produced.

The south-eastern coast is scarcely known, except towards the provinces of Haudramaut and Yemen, or Arabia Felix, where ranges of mountains, some above 5000 feet high, line the coast, and in many places project into the ocean, sometimes forming excellent harbours, as that of Aden, which is protected by jutting rocks. In the intervals there are towns and villages, cotton-trees, date-groves, and cultivated ground.

On the northern side of these granite ranges, where the table-land is 8000 feet above the sea, and along the edge of the desert of El Aklaj, in Haudramaut, there is a tract of sand so loose and so very fine, that a plummet was sunk in it by Baron Wrede to the depth of 360 feet without reaching the bottom. There is a tradition in the country that the army of King Suffi perished in attempting to cross this desert. Arabia Felix, which merits its name, is the only part of that country with permanent streams, though they are small. Here also the mountains and fertile ground run far inland, producing grain, pasture, coffee, odoriferous plants, and gums. High cliffs line the shores of the Indian Ocean and the Strait of Babelman-deb, "the Gate of Tears." The fertile country is continued a considerable way along the coast of the Red Sea, but the character of barrenness is resumed by degrees, till at length the hills and intervening terraces, on which Mecca and Medina, the holy cities of the Mahomedans, stand, are sterile wastes wherever springs do not water them. The blast of the desert, loaded with burning sand, sweeps over these parched regions. Mountains skirt the table-land to the north; and the peninsula between the Gulfs of Akaba and Suez on the Red Sea, the Eliath of Scripture, is filled by the mountain-groups of Sinai and Horeb. Jebel Housa, Mount

Sinai, on which Moses received the Ten Commandments, is 9000 feet high, surrounded by higher mountains, which are covered with snow in winter. The group of Sinai is full of springs, and verdant. At its northern extremity lies the desert of El-Teh, 70 miles long and 30 broad, in which the Israelites wandered forty years. It is covered with long ranges of high rocks, of most repulsive aspect, rent into deep clefts only a few feet wide, hemmed in by walls of rock sometimes 1000 feet high, like the deserted streets of a Cyclopean town. The whole of Arabia Petrea, Edom of the sacred writers, presents a scene of appalling desolation completely fulfilling the denunciation of prophecy.

A sandy desert, crossed by low limestone ridges, separates the table-land of Arabia from the habitable part of Syria, which the mountains of Lebanon divide into two narrow plains. These mountains may almost be considered offsets from the Taurus chain; at least they are joined to it by the wooded range of Gawoor, the ancient Amanus, impassable except by two defiles, celebrated in history as the Amanic and Syrian Gates. The group of Lebanon begins with Mount Cavius, which rises abruptly from the sea in a single peak to the height of 7000 feet, at the mouth of the river Orontes. From thence the chain runs south, at a distance of about 20 miles from the shores of the Mediterranean, in a continuous line of peaks to the sources of the Jordan, where it splits into two nearly parallel naked branches, enclosing the wide and fertile plain of Beka or Ghor, the ancient Cœlo-Syria, in which are the ruins of Balbec.

The Lebanon branch terminates at the sea near the mouth of the river Leontes, a few miles north of the city of old Tyre; while the Anti-Libanus, which begins at Mount Hermon, 9000 feet high, runs west of the Jordan through Palestine, in a winding line, till its last spurs, south of the Dead Sea, sink in ridges of rock on the desert of Sinai.

The tops of all these mountains, from Scanderoon to Jerusalem, are covered with snow in winter; it is permanent on Lebanon only, whose absolute elevation is 9300 feet. The precipices are terrific, the springs abundant, and the spurs of the mountains are studded with villages and convents; there are forests in the higher grounds, and lower down vineyards and gardens. Many offsets from the Anti-Libanus end precipitously on the coast between Tripoli and Berout, among which the scenery is superb.

The valleys and plains of Syria are full of rich vegetable mould, particularly the plain of Damascus, which is brilliantly verdant, though surrounded by deserts, the barren uniformity of which is relieved on the east by the broken columns and ruined temples of Palmyra and Tadmore. The Assyrian wilderness, however, is not everywhere absolutely barren. In the spring-time it is covered with a thin but vivid verdure, mixed with fragrant aromatic herbs, of very short duration. When these are burnt up, the unbounded plains resume their wonted dreariness. The country, high and low, becomes more barren towards the Holy Land, yet even here some of the mountains —as Carmel, Bashan, and Tabor—are luxuriantly wooded, and many of the valleys are fertile, especially the valley of the Jordan, which has the appearance of pleasure-grounds, with groves of wood and aromatic plants, but almost in a state of nature. One side of the Lake of Galilee is savage; on the other there are gentle hills and wild romantic vales, adorned with palm-trees, olives, and scycamores,—a scene of calm solitude and pastoral beauty. Jerusalem stands on a declivity encompassed by severe stony mountains, wild and desolate. The greater part of Syria is a desert compared with its ancient state. Mussulman rule has blighted this fair region, once flowing with milk and honey,—the land of promise.

Farther south desolation increases; the valleys become narrower, the hills more denuded and rugged, till south of the Dead Sea their dreary aspect announces the approach to the desert.

The valley of the Jordan affords the most remarkable instance known of the depression of the land below the general surface of the globe. This hollow, which extends from the Gulf of Accabah on the Red Sea to the bifurcation of Lebanon, is 625 feet below the level of the Mediterranean at the Sea of Galilee, and the acrid waters of the Dead Sea have a depression of 1230 feet. The lowness of the valley had been observed by the Romans, who gave it the descriptive name of Cœlo-Syria, "Hollow Syria." It is absolutely walled in by mountains between the Dead Sea and Lebanon, where it is from 10 to 15 miles wide.

A shrinking of the strata must have taken place along this coast of the Mediterranean from a sudden change of temperature, or perhaps in consequence of some of the internal props giving way, for the valley of the Jordan is not the only instance of a dip of the soil below the sea-level; the small bitter lakes on the Isthmus of Suez are cavities of the same kind, as well as the Natron lakes on the Libyan desert west from the delta of the Nile.

CHAPTER VII.

AFRICA :—TABLE-LAND—CAPE OF GOOD HOPE AND EASTERN COAST—WESTERN COAST—ABYSSINIA—SENEGAMBIA—LOW LANDS AND DESERTS.

THE continent of Africa is 5000 miles long from the Cape of Good Hope to its northern extremity, and as much between Cape Guardafui, on the Indian Ocean, and Cape Verde, on the Atlantic ; but, from the irregularity of its figure, it has an area of only 12,000,000 square miles. It is divided in two by the equator, consequently the greater part of it lies under a tropical sun. The high and low lands of this portion of the old continent are so distinctly separated by the Mountains of the Moon, that, with the exception of the mountainous territory of the Atlas, and the small table-land of Barca, it may be said to consist of two parts only, a high country and a low.

An extensive, though not very elevated table-land, occupies all southern Africa, and even reaches to six or seven degrees north of the equator. On three sides it shelves down in tiers of narrow parallel terraces to the ocean, separated by mountain-chains which rise in height as they recede from the coast; and there is reason to believe that the structure of the northern declivity is similar, though its extremities only are known—namely, Abyssinia on the east, and the high land of Senegambia on the west; both of which project farther to the north than the central part.

The borders of the table-land are very little known to Europeans, and still less its surface, which no white man has crossed north of the Tropic of Capricorn. A comparatively small part, north from the Cape of Good Hope, has been explored by European travellers. Mr. Truter and Mr. Somerville were the first white men whom the inhabitants of Litakoo had seen. Of an expedition that followed their track, a few years after, no one returned.

North of the Cape the land rises to 6000 feet above the sea, and the Orange River, with its tributaries, may be more aptly said to drain than to irrigate the arid country through which they flow; many of the tributaries, indeed, are only the channels through which torrents, from the periodical rains, are carried to the Orange River, and are destitute of water many months in the year. The "Dry River," the name of one of these periodical streams, is in that country no misnomer. Their margins are adorned with mimosas, and the sandy plains have furnished treasures to the botanist.

Dr. Smith crossed the Tropic of Capricorn in a journey from the Cape of Good Hope, where the country had still the same arid character. North from that there is a great tract unexplored. In 1802 two native travelling merchants crossed the continent, which is 1590 miles wide, from Loanda, on the Atlantic, to Zambeze, on the Mozambique Channel. They found various mercantile nations, considerably advanced in civilization, who raise abundance of maize and millet, though the greater part of the country is in a state of nature. Ridges of low hills, yielding copper, the staple commodity of this country, run from S.E. to N.W. to the west of the dominions of the Cambeze, a country full of rivers, morasses, and extensive salt-marshes, which supply this part of the continent with salt. The travellers crossed 102 rivers, most of them fordable; but the leading feature of this country is Lake N'yassi, of great but unknown length, though comparatively narrow. It begins 200 miles north from the town of Tete, on the Zambeze, and extends from S.E. to N.W., flanked on the east by a range of mountains of the same name, running in the same direction, at the distance of 350 miles from the Mozambique Channel. This is all we know of the table-land of south Africa. It is evident, however, that there can be no very high mountains coverved with perpetual snow on the table-land, for, if there were, southern Africa would not be destitute of great rivers; nevertheless, the height of the Komri, or Mountains of the Moon, on its northern edge, must be considerable, to supply the perennial sources of the Nile, the Senegal, and the Niger.

The edges of the table-land are better known. At the Cape of Good Hope the African continent is about 700 miles broad, and ends in three narrow parallel ridges of mountains, the last of which is the highest and abuts on the table-land. All are cleft by precipitous deep ravines, through which winter torrents flow to the ocean. The longitudinal valleys, or karoos, that separate them are tiers, or steps, by which the plateau dips to the maritime plains. The descent is rapid, as both these plains and the mountain-ranges are very narrow. On the western side the mountains form a high group, and end in steep promontories on the coast, where Table Mountain, at Cape Town, 3582 feet high, forms a conspicuous landmark to mariners.

Granite, which is the base of southern Africa, rises to a considerable height in many places, and is generally surmounted by vast horizontal beds of sandstone, which give that character of flatness peculiar to the summits of many of the Cape mountains.

The karoos, or longitudinal valleys, are arid deserts in the dry season, but soon after the rains they are covered with verdure and a splendid flora. The maritime plains partake of the same temporary aridity, though a large portion is rich in cereal productions, vineyards, and pasture.

The most inland of the parallel ranges, about the 20th meridian, is 10,000 feet high, and, though it sinks to some groups of hills at its eastern end, it rises again, about the 27th meridian, in a truly Alpine and continuous chain—the Quotlamba Mountains, which follow the northerly direction of Natal, and are continued in the Lupata range of hills, 80 miles inland, through Zanguebar.

At Natal the coast is grassy, with clumps of trees, like an English park. The Zambeze, and other streams from the table-land, refresh the plains on the Mozambique Channel and Zanguebar, where, though some parts are marshy and covered with mangroves, groves of palm-trees adorn the plains, which yield prodigious quantities of grain, and noble forests cover the mountains; but from 4° N. lat. to Cape Guardafui is a continued desert. There is also a barren tract at the southern end of the Lupata chain, where gold is found in masses and grains on the surface and in the water-courses, which tempted the Portuguese to make settlements on these unwholesome coasts.

The island of Madagascar, with its magnificent range of mountains, 12,000 feet high, full of tremendous precipices, and covered with primeval forests, is parallel to the African coast, and only separated from it by the Mozambique Channel; so it may be presumed that it rose from the deep at the same time as the Lupata chain.

The contrast between the eastern and western coasts of South Africa is very great. The escarped bold mountains round the Cape of Good Hope, and its rocky coast, which extends a short way along the Atlantic to the north, are succeeded by ranges of sandstone of small elevation, which separate the internal sandy desert from the equally parched sandy shore. The terraced dip of the Atlantic coast, for 900 miles between the Orange River and Cape Negro, has not a drop of fresh water.

At Cape Negro ranges of mountains, separated by long level tracts, begin, and make a semicircular bend into the interior, leaving plains along the coast 140 miles broad. In Benguela these plains are healthy and cultivated; farther north there are monotonous grassy savannahs, and forests of gigantic trees. The ground, in many places saturated with water, bears a tangled crop of mangroves and tall reeds; which even cover the shoals along that flat coast; hot pestilential vapours hang over them, never dissipated by a breeze.

The country of the Calbongos is the highest land on the coast, where a magnificent group of mountains, 13,000 feet above the sea, covered almost to their tops with large timber, lie not far inland. The low plains of Bafra and Benin, west of them, but especially the delta of the Niger, consist entirely of swamps loaded with rank vegetation. The angel of Death, brooding over these regions in noisome exhalations, guards the interior of that country from the aggressions of the European, and has hitherto baffled his attempts to form settlements on the banks of this magnificent river.

Many portions of North Guinea are so fertile that they might vie with the valley of the Nile in cereal riches, besides various other productions; and though the temperature is very high, the climate is not very unhealthy.

No European has yet seen the Mountains of the Moon, which are said to cross the continent along the northern edge of the great plateau, between the two projections or promontories of Abyssinia and Senegambia. This chain divides the semi-civilized states of Soudan, Bornou, and Begharmi from the barbarous nations on the table-land. It extends south of Abyssinia at one end, at the other it joins the high land of Senegambia, and is continued in the Kong range, which runs 1200 miles behind Dahomy and the Gold Coast, and ends in the promontory of Sierra Leone.

The vast Alpine promontory of Abyssinia, or Ethiopia, 700 miles wide, projects from the table-land for 300 miles into the low lands of North Africa. It dips in parallel ridges and longitudinal valleys to the Red Sea on the east, to a low swampy region on the north, and to the plains of Senaar and Kardofan on the west. The whole country is a mass of rugged mountains, torn by ravines, with intervening cultivated valleys and verdant plains. The plain of Dembea, the summit of the plateau, 8000 feet above the sea, the granary of the country, also abounds in pasture, and enjoys a perpetual spring. Dr. Beke, who has travelled in the south of Abyssinia to within six degrees of the equator, found the same natural characters.

The mountains of Abyssinia, and those to the west of it, are the watershed whence the streams that form the Nile flow to the north, while the Quilimane, which rises also in these mountains, runs to the Indian Ocean, and all the streams that rise east of Bornou run into Lake Tchad.

The geological structure of Abyssinia is similar to that of the Cape of Good Hope, the base being granite, and the superstructure sandstone, occasionally with limestone, schist, and breccia. The granite comes to the surface in the lower parts of Abyssinia, but sandstone predominates in the upper parts, and assumes a tabular form, often lying on the tops of the mountains in enormous flat masses, only accessible by steps cut in the rock, or by ladders. Such insulated spots are used as state prisons. Large tracts are of ancient volcanic rocks, especially in Shoa.

Senegambia, the appendage to the western extremity of the table-land, also projects far into the low lands, and is the watershed whence the streams flow on one side to the plains of Soudan, where they join the Joliba, or Niger; and from the other side, the Gambia, Senegal, and other rivers, run into the Atlantic over a rich cultivated plain, but unhealthy, from the rankness of the vegetation.

The moisture that descends from the northern edge of the table-land of South Africa, under the fiery radiance of a tropical sun, fertilizes a tract of country stretching from sea to sea across the continent, the commencement of the African low lands. A great part of this region, which contains many kingdoms and commercial cities, is a very productive country. The abundance of water, the industry of the natives in irrigating the ground, the periodical rains, and the tropical heat, leave the soil no repose. Agriculture is in a rude state, but nature is so bountiful that rice and millet are raised in sufficient quantity to supply the wants of a numerous population. Gold is found in the river-courses, and there are elephants in the forests; but man is the staple of their commerce,—a disgrace to the savage who sells his fellow-creature, but a far greater disgrace to the more savage purchaser, who dares to assume the sacred name of Christian.

This long belt of never-failing vitality, which has its large lakes, poisonous swamps, deep forests of gigantic trees, and vast solitudes in which no white man ever trode, is of small width compared with its length. In receding from the mountains the moisture becomes less, and the soil gradually worse, sufficing only to produce grass for the flocks of the wandering Beduin. At last a hideous barren waste begins, which extends northwards 800 miles in unvaried desolation to the grassy steppes at the foot of the Atlas; and, for 1000 miles between the Atlantic and the Red Sea, the nakedness of this blighted land is unbroken but by the valley of the Nile and a few oases.

In the west about 760,000 square miles, an area equal to that of the Mediterranean Sea, is covered by the trackless sands of the Sahara Desert, which is even prolonged for miles into the Atlantic in the form of sandbanks. This desert is alternately scorched by heat and pinched by cold. The wind blows from the east nine months in the year, and at the equinoxes it rushes in a hurricane, driving the sand in clouds before it, producing the darkness of night at midday, and overwhelming caravans of men and animals in common destruction. Then the sand is heaped up in waves ever varying with the blast;

even the atmosphere is of sand. The desolation of this dreary waste, boundless to the eye as the ocean, is terrific and sublime; the dry, heated air is like a red vapour, the setting sun seems to be a volcanic fire, and, at times, the burning wind of the desert is the blast of death. There are many salt-lakes to the north, and even the springs are of brine; thick incrustations of dazzling salt cover the ground, and the particles, carried aloft by whirlwinds, flash in the sun like diamonds.

Sand is not the only character of the desert; tracts of gravel and low bare rocks occur at times, not less barren and dreary; but, on the eastern and northern borders of the Sahara, fresh water rises near the surface, and produces an occasional oasis where barrenness and vitality meet. The oases are generally depressed below the level of the desert, with an arenaceous or calcareous border enclosing their emerald verdure like a frame. The smaller oases produce herbage, ferns, acacias, and some shrubs; forests of date-palms grow in the larger, which are the resort of lions, panthers, gazelles, reptiles, and a variety of birds.

In the Nubian and Libyan deserts, to the east of the Sahara, the continent shelves down towards the Mediterranean in a series of steps, consisting of vast level sandy or gravelly deserts, lying east and west, separated by low rocky ridges. This shelving country, which is only 540 feet above the sea at the distance of 750 miles inland, is cut transversely by the Nile, and by a deep furrow parallel to it, in which there is a long line of oases. This furrow, the Nile, and the Red Sea, nearly parallel to both, are flanked by rocky eminences which go north from the table-land.

On the interminable sands and rocks of these deserts no animal, no insect, breaks the dread silence; not a tree nor a shrub is to be seen in this land without a shadow. In the glare of noon the air quivers with the heat reflected from the red sand, and in the night it is chilled under a clear sky sparkling with its host of stars. Strangely, but beautifully, contrasted with these scorched solitudes is the narrow valley of the Nile, threading the desert for a thousand miles in emerald green, with its blue waters foaming in rapids among wild rocks, or quietly spreading in a calm stream amidst fields of corn, and the august monuments of past ages.

At the distance of a few days' journey west from the Nile, over a hideous flinty plain, lies the furrow already mentioned, trending to the north, and containing the oases of Darfour, Selime, the Great and Little Oases, and the parallel valleys of the Natran lakes, and Bahr-Belama, or the "Dry River." The Great Oasis, or Oasis of Thebes, is 125 miles long, and 4 or 5 broad; the Lesser Oasis, separated from it by 40 miles of desert, is of the same form. Both are rich in verdure and cultivation, with villages amid palm-groves and fruit-trees, mixed with the ruins of remote antiquity; offering scenes of peaceful and soft beauty contrasted with the surrounding gloom. The Natran lakes are in the northern part of the valley of Nitrea, 35 miles west of the Nile; the southern part is a beautiful quiet spot, that became the retreat of Christian monks in the middle of the second century, and at one time contained 360 convents, of which only four remain; from these some very valuable manuscripts of old date have recently been obtained.

Another line of oases runs along the latitude of Cairo, with fresh-water lakes, consequently no less fertile than the preceding. The ruins of the Temple of Jupiter Ammon are in one of them.

Hundreds of miles on the northern edge of the desert, from the Atlantic along the southern foot of the Atlas to the Great Syrte, are pasture-lands without a tree, an ocean of verdure. At the Great Syrte the Sahara comes to the shores of the Mediterranean, and, indeed, for 1100 miles between the termination of the Atlas and the little table-land of Barca, the ground is so unprofitable that the population only amounts to about 30,000, and these are mostly wandering tribes who feed their flocks on the grassy steppes. Magnificent countries lie along the Mediterranean coast, north of the Atlas, susceptible of cultivation. History, and the ruins of many great cities, attest their former splendour. Even now there are many populous commercial cities, and much grain is raised, though a great part of these valuable kingdoms is badly cultivated, or not cultivated at all.

The base of the sandy parts of North Africa is stiff clay. In Lower Nubia, between the parallels of Assouan and Esneh, red and white granite prevail, followed by argillaceous sandstone. Middle Egypt is calcareous, and lower down the Nile sand and alluvium cover the surface.

The prodigious extent of desert is one of the most extraordinary circumstances in the structure of the old continent. A zone of almost irretrievable desolation prevails from the Atlantic Ocean, across Africa and through Central Asia, almost to the Pacific Ocean, through at least 120 degrees of longitude. There are also many large districts of the same sterile nature in Europe; and if to these sandy plains the deserts of Siberia be added, together with all the barren and rocky mountain tracts, the unproductive land in the Old World is prodigious. The quantity of salt on the sandy plains is enormous, and proves that they have been part of the bed of the ocean, or of inland seas, at no very remote geological period. The low lands round the Black Sea and Caspian, and the Lake of Aral, seem to have been the most recently reclaimed, from the great proportion of shells in them identical with those now existing in these seas. The same may be said of the Sahara Desert, where salt and recent shells are plentiful.

CHAPTER VIII.

AMERICAN CONTINENT—THE MOUNTAINS OF SOUTH AMERICA —THE ANDES—THE MOUNTAINS OF THE PARIMA AND BRAZIL.

SOME thinner portion of the crust of the globe under the meridians that traverse the continent of America from Cape Horn to the Arctic Ocean must have yielded to the expansive forces of the subterranean fires, or been rent by the contraction of the strata in cooling. Through this the Andes had arisen, producing the greatest influence on the form of the continent, and the peculiar simplicity that prevails in its principal mountain systems, which, with very few exceptions, have a general tendency from north to south. The continent is 9000 miles long, and, its form being two great peninsulas joined by a long narrow isthmus, it is divided by nature into the three parts of South, Central, and North America; yet these three are connected by the mighty chain of the Andes, but little inferior in height to the Himalaya, running along the coast of the Pacific from within the Arctic nearly to the Antarctic circle. In this course every variety of climate is to be met with, from the rigour of polar congelation to the scorching heat of the torrid zone; while the mountains are so high that the same extremes of heat and cold may be experienced in the journey of a few hours from the burning plains of Peru to the snow-clad peaks above. In this long chain there are three distinct varieties of character, nearly, though not entirely, corresponding to the three natural divisions of the continent. The Andes of South America differ materially from those of Central America and Mexico, while both are dissimilar to the Andes of North America, generally known as the Chippewayan or Rocky Mountains.

The greatest length of South America from Cape Horn to the Isthmus of Panama is about 4550 miles. It is very narrow at its southern extremity, but increases in width northwards to the latitude of Cape Roque on the Atlantic, between which and Cape Blanco on the Pacific it attains its greatest breadth of nearly 2446 miles. It consists of three mountain systems, separated by the basins of three of the greatest rivers in the world. The Andes run along the western coast from Cape Horn to the Isthmus of Panama, in a single chain of small width but majestic height, dipping rapidly to the narrow maritime plains of the Pacific, but descending on the east in high valleys and occasional offsets to plains of vast extent, whose dead level is for hundreds of miles as unbroken as that of the ocean by which they are bounded. Nevertheless two detached mountain systems rise on these plains, one in Brazil between the Rio de la Plata and the river of the Amazons; the other is that of Parima and Guiana, lying between the river of the Amazons and the Oronoco.

The great chain of the Andes first raises its crest above the waves of the Antarctic Ocean in the majestic dark mass of Cape Horn, the southernmost point of the archipelago of Terra del Fuego. This group of mountainous islands, equal in size to Britain, is cut off from the main land by the Strait of Magellan. The islands are penetrated in every direction by bays and narrow inlets of the sea, or fiords, ending in glaciers fed by the snow on the summits of mountains 6000 feet high. Peat-mosses cover the higher declivities of these mountains, and their flanks are beset with densely entangled forests of brown beech, which never lose their dusky leaves, producing altogether a savage dismal scene. The mountains which occupy the western side of this cluster of islands sink down to wide level plains to the east, like the continent itself, of which the archipelago is but the southern extremity.

The Pacific comes to the very base of the Patagonian Andes for about 1000 miles, from Cape Horn to the 40th parallel of south latitude. The whole coast is lined by a succession of archipelagos and islands, separated from the iron-bound shores by narrow arms of the sea, which, in the more southern part, are in fact profound longitudinal valleys of the Andes filled by the ocean, so that the chain of islands running parallel to the axes of the mountains is but the tops of an exterior range rising above the sea.

The coast itself for 650 miles is begirt by walls of rock, which sink into an unfathomable depth, torn by long crevices or fiords, similar to those on the Norwegian shore, ending in tremendous glaciers, whose masses, falling with a crash like thunder, drive the sea in sweeping breakers through these chasms. The islands and the main land are thickly clothed with forests, which are of a less sombre aspect as the latitude decreases.

South of the archipelago of Chiloe there are few spots susceptible of cultivation, and none fit for the permanent habitation of man; but Chiloe itself, the most southerly part of the globe that is inhabited, is fertile. There are four magnificent volcanoes in the Andes opposite to these islands. In southern Chili the Andes retire a little from the sea, leaving plains traversed by ranges of hills 2000 or 3000 feet high, running parallel to the coast, cut by valleys and separated by flat basins, the beds of ancient lakes, now inhabited.

The Cordillera itself runs behind in a single chain, about 20 miles broad, with 12,000 feet of mean elevation. The mountain-tops maintain a horizontal line parallel to that of perpetual snow, surmounted at long intervals by groups of points, or a solitary volcanic cone, in delicate relief on the clear blue sky. Of these, Descabezado, the "Beheaded," rises 12,102 feet above the sea, and behind Valparaiso, in the centre of a knot of mountains, the magnificent volcano of Aconcagua has an absolute height of 23,000 feet. All the higher ranges of the Chilian Andes are uninhabitable; there are very few valleys which lead to the central range, and these are mostly in southern Chili; in other places the chain is utterly impassable to beasts of burthen. The flat parts of these mountains are often volcanic, and the precipices are frightful. The descent is so abrupt on both sides, that northern Chili may be esteemed a declivity of the Cordillera.

About the latitude of Conception the dense forests of semi-tropical vegetation cease with the humid equable climate; and as no rain falls in central Chili for nine months in the year, the brown, purple, and tile-red hills and mountains are only dotted here and there with low trees and bushes; very soon, however, after the heavy showers have moistened the cracked ground, it is covered with a beautiful but transient flora. In some valleys it is more permanent and of a tropical character, mixed with Alpine plants. In southern Chili rain falls only once in two or three years, the consequence of which is sterility on the western precipitous and unbroken descent of the Andes; but on the east various secondary branches leave the central Cordillera, which extend 300 or 400 miles into the plains, wooded to a great height.

The chain takes the name of the Peruvian Andes about the 24th degree of south latitude, and is separated from the Pacific for 1250 miles by a sandy desert, seldom above 60 miles broad, on which a drop of rain never falls, where bare rocks pierce through the moving sand, and which has a mine of rock-salt, a character of deserts generally. The width of the coast is nearly the same to the Isthmus of Panama, but damp luxuriant forests, full of orchideæ, begin about the latitude of Payta, and continue northwards.

From its southern extremity to the Nevada of Chorolque, in 21° 30′ S. lat., the Andes are merely a grand range of mountains, but north of that the chain becomes a very elevated narrow table-land, or longitudinal Alpine valley, in the direction of the coast, bounded on each side by a parallel row of high mountains, rising much above the table-land. These parallel Cordilleras are united at various points by enormous transverse groups or mountain-knots, or by single ranges crossing between them like dykes, a structure that prevails to Pasto in 1° 13′ 6″ N. lat. The descent to the Pacific is very steep, but the dip is less rapid to the east, whence offsets diverge to the level plains. The most remarkable peculiarity of the Andes is the absence of transverse valleys; with the exception of a few in the Patagonian and south Chilian Andes there is not an opening through these mountains in the remainder of their course to the Isthmus of Panama.

Unlike the table-lands of Asia of the same elevation, where cultivation is confined to the more sheltered spots, or those still lower in Europe, which are only fit for pasture, these lofty regions of the Andes yield exuberant crops of every European grain, and have many populous cities enjoying the luxuries of life, with universities, libraries, civil and religious establishments, at altitudes equal to that of the Peak of Teneriffe, which is 12,358 feet above the sea-level. Villages are placed and mines are wrought at heights little less than the top of Mont Blanc. This state is not limited to the present times, since these table-lands were made the centre of civilization by a race of mankind which "bear the same relation to the Incas and the present inhabitants that the Etruscans bear to the ancient Romans and to the Italians of our own days."

The table-land of Desaguadero, one of the most remarkable of these, has an absolute altitude of 13,000 feet, and a breadth varying from 30 to 60 miles: it stretches 500 miles along the top of the Andes, between the transverse mountain-group of Las Lipez, in 20° S. lat., and the enormous mountain-knot of Vilcañata and Cusco, which, extending from east to west, shuts in the valley on the north, occupying an area three times as large as Switzerland, and rising 8300 feet above the surface of the table-land, from which some idea may be formed of the gigantic scale of the Andes. This table-land or valley is bounded on each side by the two grand chains of the Bolivian Andes: that on the west is the Cordillera of the coast; the range on the east side is the Cordillera Reale. These two rows of mountains lie so near the edge that the whole breadth of the table-land, including both, is only 300 miles. All the snowy peaks of the Cordilleras of the coast, varying from 18,000 to 22,000 feet in absolute height, are either active volcanoes or of volcanic origin, and, with the exception of the volcano of Uvinas, they are all situate upon the maritime declivity of the table-land, and not more than 60 miles from the Pacific; consequently the descent is very abrupt. The eastern Cordillera, which begins at the metalliferous mountains of Pasco and Potosi, is not more than 17,000 feet high to the south, and below the level of perpetual snow, but its northern portion contains the three peaked mountains of Sorata, 25,000 feet above the sea, and is one of the most magnificent chains in the Andes. The snowy part begins with the gigantic mass of Illimani, whose serrated ridges, elongated in the direction of the axis of the Andes, rise 24,000 feet above the ocean. The lowest glacier on its southern slope does not come below 16,500 feet, and the valley of Totoral, a mere gulf 18,000 feet deep, in which Vesuvius might stand, comes between Illimani and the Nevada of Tres Cruces, from whence the Cordillera Reale runs northward in a continuous line of snow-clad peaks to the group of Vilcañata and Cusco, which unites it with the Cordilleras of the coast.

The valley or table-land of Desaguadero, occupying 150,000 square miles, has a considerable variety of surface; in the south, throughout the mining district, it is poor and cold. There Potosi, the highest city in the world, stands, at an absolute elevation of 13,350 feet, on the declivity of a mountain celebrated for its silver-mines at the height of 16,060 feet. Chiquisaca, the capital of Bolivia, containing 13,000 inhabitants, lies to the south-east of Potosi, in the midst of cultivated fields. The northern part of the valley is populous, and productive in wheat, maize, and other grain; and there is the Lake of Titicaca, twenty times as large as the Lake of Geneva. The islands and shores of this lake still exhibit ruins of gigantic magnitude, monuments of a people more ancient than the Incas. The modern city of La Paz d'Ayachuco with 40,000 inhabitants, on its southern border, stands in the most sublime situation that can be imagined, having the vast Nevada of Illimani to the north, and the no less magnificent Sorata to the south. The two ranges of the Bolivian Andes in such close approximation, with their smoking cones and serrated ridges, form one of the most august scenes in nature.

Many offsets leave the eastern side of the Cordillera Reale, which terminate in the great plain of Chiquitos and Paraguay; the most important is the Sierra Nevada de Cochobamba, which bounds a rich valley of the same name on the north, and, after dividing the basins of the Rio de la Plata from that of the Amazons for 200 miles, ends near the town of Santa Cruz de la Sierra.

There are fertile valleys and plains in the snow-capped group of Vilcañata and Cusco. The city of Cusco, which contains nearly 50,000 inhabitants, was the capital of the empire of the Incas, and the ruins of the Temple of the Sun still bear marks of its former splendour. Two ancient Peruvian roads lead from Cusco to Quito, in no respect inferior to the old Roman roads: that over the mountain plains is higher than the Peak of Teneriffe. North from Cusco lies the plain of Bombon, which assumes the bleak and dismal character common to the mining districts. It is 14,000 feet above the sea, and only 18 miles wide between the Cordilleras that bound its sides, and which send their streams into the Lake of Lauri or Laurichoco, the source of the river of the Amazons. There are many small lakes on the table-lands and high valleys of the Andes, some even within the range of perpetual snow. They are very cold and unfathomably deep, often of the purest sea-green colour, probably the craters of old volcanoes.

The crest of the Andes is split into three rows of mountains running from south to north from the transverse group of Pasco and Huanuco, which shuts in the valley of Bombon between the 11th and 10th parallels of south latitude: that in the centre separates the wide fertile valley of the upper Marañon from the still richer valley of Huallago. The western chain alone reaches the line of perpetual snow, and no mountain north of this for 400 miles to Chimborazo arrives at the snow-line.

North from the group of Loxas, celebrated for its forests of the cinchona or Peruvian bark tree, the summit of the Andes spreads into a narrow table-land, which extends 350 miles in the direction of the chain, passing through the republic of the Equator to the mountain-group of Pastos in New Grenada. It is hemmed in on each side by Cordilleras of gigantic size, and divided by the cross ridges of the Paramo del Assuay and Chisinche into three parts, namely, the plains of Cuença, Tassia, and Quito, by much the greatest. The plain of Cuença is uninteresting, but the plain of Tassia is very magnificent; the huge dome-shaped Chimborazo rises in its eastern Cordillera 21,428 feet above the sea, yet not the highest mountain in the Andes; and in the same Cordillera are the pyramidal peaks of Illiniza, the wreck of an ancient volcano. The height of Illiniza above the Pacific and above the table-land was measured by the French Academicians, and from their measurement they obtained the height of Quito, and an approximate value of the barometrical coefficient. In the western Cordillera lies the ever agitated volcano of Sangay, together with Cotopaxi, the most beautiful of volcanoes, whose cone of dazzling white is six times as high as that of the Peak of Teneriffe.

The table-land of Quito, one of the largest and finest in the Andes, is 200 miles long and 30 wide, with an absolute altitude of 10,000 feet, bounded by the most magnificent series of volcanoes and mountains in the New World. A peculiar interest is attached to two of the many magnificent volcanoes in the parallel Cordilleras that flank it on each side. In the eastern chain the beautiful snow-clad cone of Cayambe is traversed by the equator, the most remarkable division of the globe; and in the western Cordillera the cross still stands on the summit of Pinchincha, 15,924 feet above the Pacific, which served for a signal to Messieurs Bouguer and Condamine in the measurement of a degree of the meridian.

Some parts of the plain of Quito to the south are sterile, but the soil generally is good, and perpetual spring clothes it with exuberant vegetation. The city of Quito, containing 70,000 inhabitants, on the side of Pinchincha, has an absolute height of 9000 feet. The city is well built and handsome; the churches are splendid; it possesses universities, the comforts and luxuries of civilized life, in a situation of unrivalled grandeur and beauty. Thus on the very summit of the Andes there is a world by itself, with its mountains and its valleys, its lakes and rivers, populous towns and cultivated fields. Many monuments of the Incas are still found in good preservation in these plains, where the scenery is magnificent; eleven volcanoes are visible from one spot. Although the Andes are inferior in height to the Himalaya, yet the domes of trachyte, the truncated cones of the active volcanoes, and the serrated ruins of those that are extinct, mixed with the bald features of primary mountains, give an infinitely greater variety to the scene, while the smoke, and very often the flame, issuing from these regions of perpetual snow, increase its sublimity. Stupendous as these mountains appear even from the plains of the tableland, they are merely the inequalities of the tops of the Andes, the serrated summit of that mighty chain.

Between the large group of Los Pastos, containing several active volcanoes, and the group of Los Papos in the second degree of north latitude, the table-land is only 6900 feet above the sea; and north of the latter mountain-knot the crest of the Andes splits into three Cordilleras, which meet no more. The most westerly of these, the continuation of the great chain, divides the valley of the river Cauca from the Gulf of Panama; it is only 5000 feet high, and the lowest of the three. Though but 20 miles broad, it is so steep, and so difficult to pass, that travellers cannot go on mules, but are carried on men's shoulders; it is exceedingly rich in gold and platina. The central branch, or Cordillera of Quindici, runs due north between the Magdalena and Cauca, with a mean height of 10,000 feet, though rising to 18,314 feet on the Peak of Tolima. The most easterly of the three Cordilleras, called the Sierra de la Summa Paz, spreads out into the table-land of Santa Fé de Bogota, Tunja, and others, which have an elevation of about 9000 feet, and its precipices border the rivers Orinoco and Meta. The tremendous crevice of Icononza occurs in the path leading from the city of Santa Fé de Bogota to the banks of the Magdalena. It probably was formed by an earthquake, and is like an empty mineral vein, across which are two natural bridges; the lowest is composed of stones that have been jammed between the rocks in their fall. This Cordillera comprises the Andes of Cundinamarca and Merida, and goes north-east through Grenada to the 10th northern parallel, where it joins the coast-chain of Venezuela or Caraccas, which runs due east, and ends at Cape Paria in the Caribbean Sea, or rather at the eastern extremity of the island of Trinidad. This coast-chain is so majestic and beautiful that Baron Humboldt says it is like the Alps rising out of the sea without their snow. The insulated group of Santa Martha, 19,000 feet high, deeply covered with snow, stands on an extensive plain between the delta of the Magdalena and the sea-lake of Maracaybo, and is a landmark to mariners far off in the Caribbean Sea.

The passes over the Chilian Andes are numerous; that of Portilla, leading from St. Jago to Mendoza, is the highest; it crosses two ridges; the most elevated is 14,365 feet above the sea, and vegetation ceases far below its summit. Those in Peru are higher, though none reach the snow-line. In Bolivia the mean elevation of the passes in the western and eastern Cordilleras is 14,892 and 14,422 feet respectively: the peaks in the eastern Cordillera are the highest, but the passes in the western are on the most elevated part of the range, while those in the eastern are on the lowest. That leading from Sorata to the auriferous valley of Tipuani is perhaps the highest in Bolivia. From the total absence of vegetation and the intense cold it is supposed to be 16,000 feet above the Pacific; those to the north are but little lower. The pass of Quindiu in Colombia, though only 11,500 feet high, is the most difficult of all across the Andes: but those crossing the mountain-knots from one table-land to another are the most dangerous; for example, that over the

Paramo del Assuay, in the plain of Quito, where the road is nearly as high as Mont Blanc, and travellers not unfrequently perish from cold winds in attempting it.

On the western side of the Andes little or no rain falls, except at their most southern extremity, and scanty vegetation appears only in spots, or in small valleys. Excessive heat and moisture combine to cover the eastern side and its offsets with tangled forests of large trees and dense brushwood. This exuberance diminishes as the height increases, till at last the barren rocks are covered only by snow and glaciers. Nothing can surpass the desolation of these elevated regions, where nature has been shaken by terrific convulsions. The dazzling snow fatigues the eye; the huge masses of bald rock, the mural precipices, and the chasms yawning into dark unknown depths, strike the imagination; while the crash of the avalanche, or the rolling thunder of the volcano, startles the ear. In the dead of night, when the sky is clear and the wind hushed, the hollow moaning of the volcanic fire fills the Indian with superstitious dread in the deathlike stillness of these solitudes.

In the very elevated plains in the transverse groups, such as that of Bombon, however pure the sky, the landscape is lurid and colourless; the dark-blue shadows are sharply defined, and from the thinness of the air it is hardly possible to make a just estimate of distance. Changes of weather are sudden and violent; clouds of black vapour arise, and are carried by fierce winds over the barren plains; snow and hail are driven with irresistible impetuosity; and thunder-storms come on, loud and awful, without warning. Notwithstanding the thinness of the air, the crash of the peals is quite appalling, while the lightning runs along the scorched grass, and sometimes, issuing from the ground, destroys a team of mules or a flock of sheep at one flash.*

Currents of warm air are occasionally met with on the crest of the Andes—an extraordinary phenomenon in such gelid heights, which is not yet accounted for: they generally occur two hours after sunset, are local and narrow, not exceeding a few fathoms in width; similar to the equally partial blasts of hot air in the Alps. A singular instance, probably of earth-light, occurs in crossing the Andes from Chili to Mendoza: on this rocky scene a peculiar brightness occasionally rests, a kind of undescribable reddish light, which vanishes during the winter rains, and is not perceptible on sunny days. Dr. Pœppig ascribes the phenomenon to the dryness of the air; he was confirmed in his opinion from afterwards observing a similar brightness on the coast of Peru, and it has also been seen in Egypt.

The Andes descend to the eastern plains by a series of cultivated levels, as those of Tucuman, Salta, and Jujuy, in the republic of La Plata, with many others. That of Tucuman is 3600 feet above the sea, the garden of the republic.

* Dr. Pœppig's 'Travels in South America.'

The low lands to the east of the Andes are divided by the table-lands and mountains of Parima and Brazil into three parts, of very different aspect —the deserts and pampas of Patagonia and Buenos Ayres; the Silvas, or woody basin of the Amazons; and the Llanos, or grassy steppes of the Orinoco. The eastern table-lands nowhere exceed 2500 feet of absolute height; and the plains are so low and flat, especially at the foot of the Andes, that a rise of 1000 feet in the Atlantic Ocean would submerge more than half the continent of South America.

The system of the Parima is a group of mountains scattered over a table-land not more than 2000 feet above the sea, which extends 600 or 700 miles from east to west, between the river Orinoco, the Rio Negro, the Amazons, and the Atlantic Ocean. It is quite unconnected with the Andes, being 80 leagues east from the mountains of New Grenada. It begins 60 or 70 miles from the coast of Venezuela, and ascends by four successive terraces to undulating plains which come within one or two degrees of the equator, and is twice as long as it is broad.

Seven chains, besides groups of mountains, cross the table-land from west to east, of which the chief is the Sierra del Parima. Beginning at the mouth of the Meta, it crosses the plains of Esmeralda to the borders of Brazil, whence, under the name of the Sierra Pacaraime, it goes to the left bank of the Rupuniri, a tributary of the Essequibo; then, bending to the south, it runs in a tortuous line between Brazil and Guiana to the Atlantic. This chain, not more than 600 feet high, is everywhere escarped, and forms the watershed between the tributaries of the Amazons and those of the Orinoco, the Essequibo, and the rivers of Guiana. The Orinoco rises on the northern side of the Sierra del Parima, and in its circuitous course over the plains of Esmeralda it breaks through the western extremity of that chain in two places, 12 leagues asunder, where it dashes with violence against the transverse shelving rocks and dykes, forming the splendid series of rapids and cataracts of Maypures and Atures, from whence the Parima Mountains have got the name of the Cordillera of the cataracts of the Orinoco. The chain is of granite, which forms the banks and fills the bed of the river, covered with luxuriant tropical vegetation, especially palm-forests. In the district of the Upper Orinoco, near Charichana, there is a granite rock which emits musical sounds at sunrise, like the notes of an organ, occasioned by the difference of temperature of the external air and that which fills the deep narrow crevices with which the rock is everywhere torn. Something of the same kind occurs at Mount Sinai.

The other parallel chains that extend over the table-land in Venezuela and Guiana are separated by flat savannahs, generally barren in the dry season, but after the rains covered with a carpet of emerald-green grass, often six feet high, mixed with flowers. The vegetation in these countries is splendid beyond imagination: the regions of the Upper Orinoco and Rio Negro, and of almost all the mountains and banks of rivers in Guiana, are clothed with majestic and impenetrable forests, whose moist and hot recesses are the abode of the singular and beautiful race of the Orchideæ and tangled creepers of many kinds.

Although all the mountains of the system of Parima are wild and rugged, they are not high; the inaccessible peak of the Cerro Duida, which rises insulated 7155 feet above the plain of Esmeralda, is the culminating point, and the highest mountain in South America east of the Andes.

The fine savannahs of the Rupununi were the country of romance in the days of Queen Elizabeth. South of the Pacaraime, near an inlet of the river, the far-famed city of Manoa was supposed to stand, the object of the unfortunate expedition of Sir Walter Raleigh; about 11 miles south-west of which is situated the Lake Amucu, "the Great Lake with golden banks,"—great only during the periodical floods.

On the southern side of the basin of the river Amazons lies the table-land of Brazil, nowhere more than 2500 feet high, which occupies half of that empire, together with part of the Argentine Republic and Uruguay Orientale. Its form is a triangle, whose apex is at the confluence of the rivers Marmora and Beni, and its base extends, near the shore of the Atlantic, from the mouth of the Rio de la Plata to within three degrees of the equator. It is difficult to define the limits of this vast territory, but some idea may be formed of it by following the direction of the rapids and cataracts of the rivers descending from it to the plains around. Thus a line drawn from the fall of the river Tocantines, in 3° 30′ S. lat., to the cataracts of the Madiera, in the eighth degree of south latitude, will nearly mark its northern boundary: from thence the line would run S.W. to the junction of the Marmora and Beni; then, turning to the S.E. along the Serro dos Paricis, it would proceed south to the cataract of the Paranà, called the Sete Quedas, in 24° 30′ S. lat.; and lastly, from thence, by the great falls of the river Iguassu, to the Morro de Santa Martha, at the mouth of the Rio de la Plata.

Chains of mountains, nearly parallel, extend from south-west to north-east, 700 miles along the base of the triangle, with a breadth of about 400 miles. Of these the Sierra do Mar, or the "coast-chain," reaches from the river Uruguay to Cape San Roque, never more distant than 20 miles from the Atlantic, except to the south of the bay of Santos, where it is 80. Offsets diverge to the right and left: the granite peak of Corcovedo, in the bay of Rio de Janeiro, 2306 feet high, is the end of one. The parallel chain of Espenhaço, beginning near the town of San Paolo, and forming the western boundary of the basin of the Rio San Francisco, is the highest in Brazil, one of its mountains being 8426 feet above the sea. All the mountains in Brazil have a general tendency from S.W. to N.E., except the transverse chain of Sierra das Vertentes, which begins 60 miles south of Villa Rica, and runs in a tortuous line to its termination near the junction of the Marmora and Beni, in 11° S. lat. It forms the watershed of the tributaries of the San Francesco and Amazons on the north, and those of the Rio de la Plata on the south; its greatest height is 3500 feet above the sea, but its western part, the Sierra Paricis, is merely a succession of detached hills. This chain, the coast-chain of Venezuela and the mountains of Parima, are the only ranges on the continent of America that do not entirely, or in some degree, lie in the direction of the meridians.

Magnificent forests of tall trees, bound together by tangled creeping and parasitical plants, clothe the declivities of the mountains, and line the borders of the Brazilian rivers, where the soil is rich and the verdure brilliant. Many of the plains on the table-land bear a coarse nutritious grass after the rains only; but vast undulating tracts are always verdant with excellent pasture, intermixed with fields of corn: some parts are bare sand and rolled quartz; and the Campas Paricis, north of the Sierra Vestentes, in Matto Grasso, is a sandy desert of unknown extent, similar to the Great Gobi on the table-land of Tibet.

CHAPTER IX.

THE LOW LANDS OF SOUTH AMERICA—DESERT OF PATAGONIA —THE PAMPAS OF BUENOS AYRES—THE SILVAS OF THE AMAZONS—THE LLANOS OF THE ORINOCO AND VENEZUELA —GEOLOGICAL NOTICE.

THE southern plains are the most barren of the three great tracts of American low lands; they stretch from Terra del Fuego over 27 degrees of latitude, or 1900 miles, nearly to Tucuman and the mountains of Brazil. Palms grow at one end, deep snow covers the other many months in the year. This enormous plain, of 1,620,000 square miles, begins on the eastern part of Terra del Fuego, which is a flat covered with trees, and therefore superior to its continuation on the continent through eastern Patagonia, which, for 800 miles from the land's end to beyond the Rio Colorado, is a desert of shingle. It is occasionally diversified by huge boulders, tufts of brown grass, low bushes armed with spines, brine lakes, incrustations of salt white as snow, and by black basaltic platforms, like plains of iron, at the foot of the Andes, barren as the rest. Eastern Patagonia, however, is not one universal flat, but a succession of shingly horizontal plains at higher and higher levels, separated by long lines of cliffs or escarpments, the gable ends of the tiers or plains. The ascent is small, for even at the foot of the Andes the highest of these platforms is only 3000 feet above the ocean. The plains are here and there intersected by a ravine or a stream, the waters of which do not fertilize the blighted soil. The transition from intense heat to intense cold is rapid, and piercing winds often rush in hurricanes over these deserts, shunned even by the Indian, except when he crosses them to visit the tombs of his fathers. The shingle ends a few miles to the north of the Rio Colorado: there the red calcareous earth of the Pampas begins, monotonously covered with coarse tufted grass, without a tree or bush. This country, nearly as level as the sea, and without a stone, extends almost to the table-land of Brazil, and for 1000 miles between the Atlantic and the Andes, interrupted only at vast distances by a solitary umbú, the only tree of this soil, rising like a great landmark. This wide space, though almost destitute of water, is not all of the same description. In the Pampas of Buenos Ayres there are four distinct regions. For 180 miles west from Buenos Ayres they are covered with thistles and lucern of the most vivid green so long as the moisture from the rain lasts. In spring the verdure fades, and a month afterwards the thistles shoot up 10 feet high, so dense and so protected by spines that they are impenetrable. During summer the dried stalks are broken by the wind, and the lucern again spreads freshness over the ground. The Pampas for 430 miles west of this region is a thicket of long tufted luxuriant grass, intermixed with gaudy flowers, affording inexhaustible pasture to thousands of horses and cattle; this is followed by a tract of swamps and bogs, to which succeeds a region of ravines and stones, and, lastly, a zone, reaching to the Andes, of thorny bushes and dwarf trees in one dense thicket. The flat plains in Entre Rios in Uruguay, those of Santa Fé, and a great part of Cordova and Tucuman, are of sward, with cattle-farms. The banks of the Paraná, and other tributaries of the La Plata, are adorned with an infinite variety of tropical productions, especially the graceful tribe of palms; and the river islands are bright with orange-groves. A desert of sand, called Il Gran Chaco, exists west of the Paraguay, the vegetable produce of which is confined to varieties of the aloe and cactus tribes, the last the food of the cochineal insect, which forms a valuable article of commerce. Adjoining this desert are the unknown regions of the Chiquitos and Moxos, covered with forests and jungle.

The Pampas of Buenos Ayres, 1000 feet above the sea, sinks to its level along the foot of the Andes, where the streams from the mountains collect in large lakes, swamps, lagoons of prodigious size, and wide-spreading salines. The swamp or lagoon of Ybera, of 1000 square miles, is entirely covered with aquatic plants. These swamps are swollen to thousands of square miles by the annual floods of the rivers, which also inundate the Pampas, leaving a fertilizing coat of mud. Multitudes of animals perish in the floods, and the drought that sometimes succeeds is more fatal. Between the years 1830 and 1832 two millions of cattle died from want of food. Millions of animals are sometimes destroyed by casual and dreadful conflagrations in these countries, when covered with dry grass and thistles.

The Silvas of the river of the Amazons, lying in the centre of the continent, form the second division of the South American low lands. This country is more uneven than the Pampas, and the vegetation is so dense that it can only be penetrated by sailing up the river or its tributaries. The forests not only cover the basin of the Amazons, but also its limiting mountain-chains, the Sierra Vertentes and Parima; so that the whole forms an area of woodland more than six times the size of France, lying between the eighteenth parallel of south latitude and the seventh of north; consequently intertropical and traversed by the equator. There are some marshy savannahs between the third and fourth degrees of north latitude, and some grassy steppes south of the Pacaraim chain; but they are insignificant compared with the Silvas, which extend 1500 miles along the river, varying in breadth from 350 to 800 miles, and probably more. According to Baron Humboldt, the soil, enriched for ages by the spoils of the forest, consists of the richest mould. The heat is suffocating in the deep and dark recesses of these primeval woods, where not a breath of air penetrates, and where, after being drenched by the periodical rains, the damp is so excessive that a blue mist rises in the early morning among the huge stems of the trees, and envelops the entangled creepers stretching from bough to bough. A deathlike stillness prevails from sunrise to sunset; then the thousands of animals that inhabit these forests join in one loud discordant roar, not continuous, but in bursts. The beasts seem to be periodically and unanimously roused, by some unknown impulse, till the forest rings in universal uproar. Profound silence prevails at midnight, which is broken at the dawn of morning by another general roar of the wild chorus. Nightingales, too, have their fits of silence and song: after a pause, they

> "—— all burst forth in choral minstrelsy,
> As if some sudden gale had swept at once
> A hundred airy harps."*

The whole forest often resounds, when the animals, startled from their sleep, scream in terror at the noise made by bands of its inhabitants flying from some night-prowling foe. Their anxiety and terror before a thunder-storm is excessive, and all nature seems to partake in the dread. The tops of the lofty trees rustle ominously, though not a breath of air agitates them; a hollow whistling in the high regions of the atmosphere comes as a warning from the black floating vapour; midnight darkness envelops the ancient forests, which soon after groan and creak with the blast of the hurricane. The gloom is rendered still more hideous by the vivid lightning and the stunning crash of thunder. Even fishes are affected with the general consternation; for in a

* Wordsworth.

few minutes the Amazons rages in waves like a stormy sea.

The Llanos of the Orinoco and Venezuela, covered with long grass, form the third department of South American low lands, and occupy 153,000 square miles between the deltas of the Orinoco and the river Coqueta, flat as the surface of the sea; frequently there is not an eminence a foot high in 270 square miles. They are twice as long as they are broad; and, as the wind blows constantly from the east, the climate is the more ardent the farther west. These steppes for the most part are destitute of trees or bushes, yet in some places they are dotted with the mauritia and other palm-trees. Flat as these plains are, there are in some places two kinds of inequalities: one consists of banks or shoals of grit or compact limestone, five or six feet high, perfectly level for several leagues, and imperceptible except on their edges; the other inequality can only be detected by the barometer or levelling instruments; it is called a Mesa, and is an eminence rising imperceptibly to the height of some fathoms. Small as the elevation is, a mesa forms the watershed, from S.W. to N.E., between the affluents of the Orinoco and the streams flowing to the northern coast of Terra Firma. In the wet season, from April to the end of October, the tropical rains pour down in torrents, and hundreds of square miles of the Llanos are inundated by the floods of the rivers. The water is sometimes 12 feet deep in the hollows, in which so many horses and other animals perish that the ground smells of musk, an odour peculiar to many South American quadrupeds. From the flatness of the country, too, the waters of some affluents of the Orinoco are driven backwards by the floods of that river, especially when aided by the wind, and form temporary lakes. When the waters subside these steppes, manured by the sediment, are mantled with verdure, and produce ananas with occasional groups of palm-trees, and mimosas skirt the rivers. When the dry weather returns, the grass is burnt to powder, the air is filled with dust raised by currents occasioned by difference of temperature, even where there is no wind. If by any accident a spark of fire falls on the scorched plains, a conflagration spreads from river to river, destroying every animal, and leaves the clayey soil sterile for years, till vicissitudes of weather crumble the brick-like surface into earth.

The Llanos lie between the equator and the Tropic of Cancer; consequently the mean annual temperature is about 84° of Fahrenheit. The heat is most intense during the rainy season, when tremendous thunder-storms are of common occurrence.

GEOLOGY OF SOUTH AMERICA.

The most remarkable circumstance in the geological arrangement of South America is the vast but partial development of volcanic force, which is confined to the chain of the Andes, and even in some parts only to the western Cordillera, while not a trace of it is to be found either on the great plains to the east, or on the table-lands which divide them. The actual vents occur in linear groups. The most southern of these extends from Yntales in Patagonia to the volcanoes of central Chili, a distance of 800 miles: the second volcanic line, occupying 600 miles of latitude, lies between Araquipo and Patas:* the third extends over 300 miles between Riobamba and Popayan. That these groups of active volcanoes are connected there can be little doubt, as they are only separated by a few hundred miles; and thus there is a line of volcanic action, 1700 miles long, entirely confined to the Andes, to which the volcanic islands of Juan Fernandez and the Galapagos form a parallel line.

Granite, which seems to be the base of the whole continent, is widely spread to the east and south: it appears in Terra del Fuego and in the Patagonian Andes abundantly and at great elevations; but it comes into view so rarely in the other parts of the chain that Baron Humboldt says a person might travel years in the Andes of Peru and Quito without falling in with it: he never saw it at a greater height above the sea than 11,500 feet. Gneiss is here and there associated with the granite, but mica-schist is by much the most common of the crystalline rocks. Quartz is also much developed, generally mixed with mica, and rich in gold, mercury, specular iron, and sulphur. It sometimes extends several leagues in the western declivities of Peru, 6000 feet thick. Red

* Mr. Darwin.

sandstone, of vast dimensions, and of different geological periods, occurs in the Andes, and on the tableland east of them, where in some places, as in Colombia, it spreads over thousands of miles to the shores of the Atlantic. It is widely extended at altitudes of 10,000 and 12,000 feet; for example, on the table-lands of Tarqui and Cuença. Coal is sometimes associated with it, and is found at Huenca in Peru 14,750 feet above the sea.

Porphyry abounds all over the Andes, from Patagonia to Colombia, at every elevation, on the slopes and summits of the mountains, sometimes 19,000 feet thick, but not uniformly of the same age or nature. The variety of most frequent occurrence is rich in metals, while another is destitute of them. The bare and precipitous porphyry rocks give great variety to the colouring of the Andes, especially in Chili, where purple, tile-red, and brown are contrasted with the snow on the summit of the chain.*

Trachyte is almost as abundant as porphyry. Many of the loftiest parts and all the great dome-shaped mountains in the Andes are formed of it. Masses of this rock, from 14,000 to 18,000 feet thick, are seen in Chimborazo and Pinchincha. Prodigious quantities of volcanic products, lava, tufa, and obsidian, occur on the western face of the Andes, where volcanoes are active. On the eastern side there are none. This is especially the case in that part of the chain lying between the equator and Chili. The

* Dr. Pœppig.

Bolivian Cordilleras, which are the boundary of the valley of Desaguadero, furnish a striking example. The Cordillera of the coast is entirely composed of obsidian, trachyte, and tufa, while the eastern Cordillera consists of syenite, mica-schist, porphyry, and sandstone; marl, containing gypsum, oolitic limestone, and rock salt, of the most beautiful colours. Towards Chili and throughout the Chilian range the case is different, because active volcanoes are there in the centre of the chain.

Sea-shells of different geological periods are found at various elevations, which shows that many upheavings and subsidences have taken place in the chain of the Andes, especially at its southern extremity. Stems of large trees, which Mr. Darwin found in a fossil state in the Upsallata range, a collateral branch of the Chilian Andes, now 700 miles distant from the Atlantic, exhibit a remarkable example of such vicissitudes. These trees, with the volcanic soil on which they had grown, had sunk from the beach to the bottom of a deep ocean, from which, after five alternations of sedimentary deposits and deluges of submarine lava of prodigious thickness, the whole mass was raised up, and now forms the Upsallata chain. Subsequently, by the wearing of streams, the embedded trunks have been brought into view in a silicified state, projecting from the soil on which they grew—now solid rock.

"Vast and scarcely comprehensible as such changes must ever appear, yet they have all occurred within a period recent when compared with the history of the Cordillera; and the Cordillera itself is absolutely modern, compared with many of the fossiliferous strata of Europe and America."*

From the quantity of shingle and sand on the valleys in the lower ridges, as well as at altitudes from 7000 to 9000 feet above the present level of the sea, it appears that the whole area of the Chilian Andes has been raised by a gradual motion; and the coast is now rising by the same imperceptible degrees, though it is sometimes suddenly elevated by a succession of small upheavings of a few feet by earthquakes, similar to that which shook the continent for a thousand miles on the 20th of February, 1835.

On the eastern side of the Andes the land from Tera del Fuego to the Rio de la Plata has been raised *en masse* by one great elevating force, acting equally and imperceptibly for 2000 miles, within the period of the shell-fish now existing, which in many parts of these plains even still retain their colours. The gradual upward movement was interrupted by at least eight long periods of rest, marked by the edges of the successive plains, which, extending from south to north, had formed so many lines of sea-coast, as they rose higher and higher between the Atlantic and the Andes. It appears, from the shingle and fossil shells found on both sides of the Cordillera,

* Darwin's Journal of Travels in South America.

that the whole south-western extremity of the continent has been rising slowly for a long time, and indeed the whole Andean chain.

The instability of the southern part of the continent is less astonishing if it be considered that at the time of the earthquake of 1835 the volcanoes in the Chilian Andes were in eruption contemporaneously for 720 miles in one direction and 400 in another; so that in all probability there was a subterranean lake of burning lava below this end of the continent twice as large as the Black Sea.*

The terraced plains of Patagonia, which extend hundreds of miles along the coast, are tertiary strata, not in basins, but in one great deposit, above which lies a thick stratum of white pumaceous substance, extending at least 500 miles, a tenth part of which consists of marine infusoria. Over the whole lies the shingle already mentioned, spread over the coast for 700 miles in length, with a mean breadth of 200 miles, and 50 feet thick. These myriads of pebbles, chiefly of porphyry, have been torn from the rocks of the Andes, and water-worn, at a period subsequent to the deposition of the tertiary strata—a period of incalculable duration. All the plains of Terra del Fuego and Patagonia, on both sides of the Andes, are strewed with huge boulders, transported by icebergs, which had descended to lower latitudes in ancient times than they

* Darwin's Journal of Travels in South America.

do now—observations of great interest, which we owe to Mr. Darwin.

The stunted vegetation of these sterile plains was sufficient to nourish large animals of the pachydermata tribe, now extinct, even at a period when the present shell-fish of the Patagonian seas existed.

The Pampas of Buenos Ayres are entirely alluvial, the deposit of the Rio de la Plata. Granite prevails to the extent of 2000 miles along the coast of Brazil, and with syenite forms the base of the table-land. The superstructure of the latter consists of metamorphic and old igneous rocks, sandstone, clay-slate, limestone, in which are large caverns with bones of extinct animals, and alluvial soil. Gold is found in the channels of the rivers, and no country is so rich in diamonds.

The fertile soil of the Silvas has travelled from afar. Washed down from the Andes, it has been gradually deposited and manured by the decay of a thousand forests. Granite again appears in more than its usual ruggedness in the table-land and mountains of the Parima system. The sandstone of the Andes is found there also in a chain 7300 feet high; and on the plains of Esmeralda it caps the granite of the solitary prism-shaped Duido, the culminating mountain of the Parima system. Limestone appears in the Brigantine or Cocallar, the most southern of the three ranges of the coast-chain of Venezuela; the other two are of granite, metamorphic rocks, and crystalline schists, torn by earthquakes and worn by the sea, which has deeply indented that coast. The chain of islands in the Spanish main is merely the wreck of a more northern ridge, broken up into detached masses by these irresistible powers.

CHAPTER X.

CENTRAL AMERICA—WEST INDIAN ISLANDS—GEOLOGICAL NOTICE.

TAKING the natural divisions of the continent alone into consideration, Central America may be regarded as lying between the 7th and 20th parallels of north latitude, and consequently in a tropical climate. The narrow tortuous strip of land which unites the continents of North and South America stretches from S.E. to N.W. about 1000 miles, varying in breadth from 30 miles to 300 or 400.

As a regular chain, the Andes terminate suddenly at the plain of Panama, but as a mass of high land they continue through Central America and Mexico, in an irregular mixture of table-lands and mountains. These table-lands, however, differ from those in the Andes of South America, inasmuch as they are not bounded on each side by Cordilleras following the direction of the chain, but are traversed by ranges running over them in all directions, or studded by mountains. The mass of high land which forms the central ridge of the country, and the watershed between the two oceans, is very steep on its western side, and runs near the coast of the Pacific, where Central America is narrow; but to the north, where it becomes wider, the high land recedes to a greater distance from the shore than the Andes do in any other part between Cape Horn and Mexico.

The plains of Panama, very little raised above the sea, but in some parts studded with hills, follow the direction of the isthmus for 280 miles, and end at the Bay of Parita. From thence a mass, about 3000 feet high, of forest-covered table-lands and complicated mountains extends through Veragua and Porta Rica to the Lake of Nicaragua. The plain of Nicaragua, together with its lake, occupies an area of 30,000 square miles, and forms the second break in the great Andean chain. The lake is only 128 feet above the Pacific, from which it is separated by a line of active volcanoes. The river San Juan de Nicaragua flows from its eastern end into the Caribbean Sea, and its northern extremity is connected with the smaller lake of Managua by the river Panalaya. By this water-line it has been projected to unite the two seas. The high land begins again, after an interval of 170 miles, with the Mosquito country and Honduras, which mostly consist of table-lands, high mountains, and some volcanoes.

The broad elevated belt of Guatemala lies between the Isthmus of Chiquimala and that of Tehuantepec. It spreads out to the east and forms the high but narrow table-land on the peninsula of Yucatan, which terminates at Cape Catoch, and which is bounded by high mountains and terraces along the Gulf of Honduras. The table-land of Guatimala consists of undulating verdant plains of great extent, of the absolute height of 5000 feet, fragrant with flowers. In the southern part of the table-land the cities of Old and New Guatemala are situate, 12 miles apart. The portion of the plain on which the new city stands is bounded on the west by the three volcanoes of Pacaya, del Fuego, and d'Agua; these, rising from 7000 to 10,000 feet above the plain, lie close to the new city on the west, and form a scene of wonderful boldness and beauty. The Volcano de Agua, at the foot of which Old Guatemala stands, is a perfect cone, verdant to its summit, which occasionally pours forth torrents of boiling water and stones. The old city has been twice destroyed by it, and is now nearly deserted on account of violent earthquakes. The Volcano del Fuego generally emits smoke from one of its peaks, and the volcano de Pacayo is only occasionally active. The wide grassy plains are cut by deep valleys to the north, where the high land of Guatemala ends in parallel ridges of mountains, called the Cerro Pelado, which run from east to west along the 94th meridian, filling half the Isthmus of Tehuantepec, which is 140 miles broad, and unites the table-land of Guatemala with that of Mexico.

Though there are large savannahs on the high plains of Guatemala, there are also magnificent primeval forests, as the name of the country implies, Guatemala, in the Mexican language, signifying a place covered with trees. The banks of the Rio de la Papian, or Usumasinta, which rises in the Alpine lake of Lacandon and flows over the table-land to the Gulf of Mexico, are beautiful beyond description.

The coasts of Central America are generally narrow, and in some places the mountains and high lands come close to the water's edge. The sugar-cane is indigenous, and on the low lands of the eastern coast all the ordinary produce of the West Indian Islands is raised, besides much that is peculiar to the country.

As the climate is cool on the high lands, the vegetation of the temperate zone is in perfection. On the low lands, as in other countries where heat and moisture are in excess, and where nature is for the most part undisturbed, vegetation is vigorous to rankness; forests of gigantic timber seek the free air above an impenetrable undergrowth, and the mouths of the rivers are dense masses of jungle with mangroves, and reeds 100 feet high: yet delightful savannahs vary the scene, and wooded mountains dip into the water.

Nearly all the coast of the Pacific is skirted by an alluvial plain, of small width, and generally very different in character from that on the Atlantic side. In a line along the western side of the table-land and the mountains, there is a continued succession of volcanos, at various distances from the shore, and at various heights, on the declivity of the table-land. It seems as if a great crack or fissure had been produced in the earth's surface, along the junction of the mountains and the shore, through which the internal fire had found a vent. There are more than 20 active volcanos in succession, between the 10th and 20th parallels of north latitude, some higher than the mountains of the central ridge, and several subject to violent eruptions.

The Colombian Archipelago, or West Indian Islands, which may be regarded as the wreck of a submerged part of the continent of South and Central America, consists of three distinct groups, namely, the Lesser Antilles, or Caribbean Islands, the Greater Antilles, and the Bahama or Lucay Islands. Some of the Lesser Antilles are flat, but their general character is bold, with a single mountain or group of mountains in the centre, which slopes to the sea all around, more precipitously on the eastern side, which is exposed to the force of the Atlantic current. Trinidad is the most southerly of a line of magnificent islands, which form a semicircle, enclosing the Caribbean Sea, with its convexity facing the east. The row is single to the island of Guadaloup, where it splits into two chains, known as the Windward and Leeward Islands. Trinidad, Tobago, St. Lucia, and Dominica, are particularly mountainous, and the mountains are cut by deep narrow ravines, or gullies, covered by ancient forests. The volcanic islands, which are mostly in the single part of the chain, have conical mountains bristled with rocks of a still more rugged form; but almost all the islands of the Lesser Antilles have a large portion of excellent vegetable soil in a high state of cultivation. Most of them are surrounded by coral reefs, which render navigation dangerous, and there is little intercourse between these islands, and still less with the Greater Antilles, on account of the prevailing winds and currents, which make it difficult to return. The Lesser Antilles terminate with the group of the Virgin Islands, which are small and flat, some only a few feet above the sea, and most of them are mere coral rocks.

The four islands which form the group of the Greater Antilles, are the largest and finest in the archipelago. Porto Rico, Haiti, and Jamaica, separated from the Virgin Islands by a narrow channel, lie in a line parallel to the coast-chain of Venezuela, from east to west; while Cuba, by a serpentine bend, separates the Caribbean Sea, or Sea of the Antilles, rom the Gulf of Mexico. Porto Rico is 140 miles long and 36 broad, with wooded mountains passing through its centre nearly from east to west, which furnish abundance of water. There are extensive savannahs in the interior, and very rich soil on the northern coast, but the climate is unhealthy.

Haiti, 450 miles long and 110 broad, has a group of mountains in its centre, the highest of which is 9000 feet above the sea. Chains diverge from this nucleus to the remotest parts of the island, so that there is a great proportion of high land. The mountains are susceptible of cultivation nearly to the summits, and they are clothed with undisturbed tropical forests. The extensive plains are well watered, and the soil though not deep is productive.

Jamaica, the most valuable of the British possessions in the West Indies, has an area of 4256 square miles, of which 110,000 acres are cultivated chiefly as sugar-plantations. The principal chain of the Blue Mountains lies in the centre of the island, from east to west, 5000 or 6000 feet above the sea, with so sharp a crest that in some places it is only four yards across. The offsets from it cover all the eastern part of the island; some of them are 7000 feet high. The more elevated ridges are flanked by lower ranges, descending to verdant savannahs. The escarpments are wild, the declivities steep, and mingled with stately forests. The valleys are very narrow, and not more than a twentieth part of the island is level ground. There are many small rivers, and the coast-line is 500 miles long, with at least 30 good harbours. The mean summer-heat is 80° of Fahrenheit, and that of winter 75°. The plains are often unhealthy, but the air on the mountains is salubrious; fever has never prevailed at the elevation of 2500 feet.

Cuba, the largest island in the Colombian Archipelago, has an area of 42,212 square miles, and 200 miles of coast, but so beset with coral reefs, sandbanks and rocks, that only a third of it is accessible. Its mountains, which attain the height of 8000 feet, occupy the centre, and fill the eastern part of the island, in a great longitudinal line. No island in these seas is more important with regard to situation and natural productions; and although much of the low ground is swampy and unhealthy, there are vast savannahs, and about a seventh part of the island is cultivated.

The Bahama Islands are the least valuable and least interesting part of the Archipelago. The group consists of about 500 islands, many of them mere rocks, lying east from Cuba and the coast of Florida. Twelve are rather large, and cultivated; and though arid, they produce Campeche wood and mahogany. The most intricate labyrinth of shoals and reefs, chiefly of corals, madrepores, and sand, encompass these islands; some of them rise to the surface, and are adorned with groves of palm-trees. The Great Bahama Island is the first part of the New World on which Columbus landed; the next was Haiti, where his ashes rest.

The geology of Central America is little known; nevertheless it appears, from the confused mixture of table-lands and mountain-chains in all directions, that the subterraneous forces must have acted more partially and irregularly than either in South or North America. Granite, gneiss, and mica-slate form the substrata of the country; but the abundance of igneous rocks bears witness to strong volcanic action, both in ancient and in modern times, which still maintains its activity in the volcanic groups of Guatemala and Mexico.

From the identity of the fossil remains of extinct quadrupeds, there is every reason to believe that the West Indian Archipelago was once part of South America, and that the rugged and tortuous isthmus of Central America, and the serpentine chain of islands winding from Cumana to the peninsula of Florida, are but the shattered remains of an unbroken continent. The powerful volcanic action in Central America and Mexico, the volcanic nature of many of the West Indian Islands, and the still-existing fire in St. Vincent's, together with the tremendous earthquakes to which the whole region is subject, render it more than probable that the Caribbean Sea and the Gulf of Mexico are one great area of subsidence, which possibly has been increased by the erosion of the Gulf-stream and ground-swell—a temporary current of great impetuosity, common among the West Indian Islands from October to May.

The subsidence of this extensive area must have been very great, since the water is of profound depth between the islands, and it must have taken place after the destruction of the great quadrupeds, and consequently at a very recent geological period. The elevation of the table-land of Mexico may have been a contemporaneous event. The action in the Colombian Archipelago is now, however, in a contrary direction, as the bed of the ocean is rising there. The line of volcanic islands begins with St. Vincent's, and ends with Guadaloup; the island of St. Eustasius in the Leeward range is also volcanic. The Windward and Bahama Islands are of calcareous and coral rocks. The Greater Antilles are both crystalline and calcareous in their principal mountain-chains, which are all parallel to the great chain of Venezuela, with the exception of Cuba, where the mountains diverge from a central nucleus to its extremities: there is a region of serpentine, rich in minerals, in one part of the island, with an extensive formation of columnar white marble adjacent to it.

CHAPTER XI.

NORTH AMERICA—TABLE-LAND AND MOUNTAINS OF MEXICO —THE ROCKY MOUNTAINS—THE MARITIME CHAIN AND MOUNTAINS OF RUSSIAN AMERICA.

ACCORDING to the natural division of the continent, North America begins about the 20th degree of north latitude, and terminates in the Arctic Ocean. It is longer than South America, but the irregularity of its outline renders it impossible to estimate its area. Its greatest length is about 3100 miles, and its breadth, at the widest part, is 3500 miles.

The general structure of North America is still more simple than that of the southern part of the continent. The table-land of Mexico and the Rocky Mountains, which are the continuation of the high land of the Andes, run along the western side, but at a greater distance from the Pacific; and the immense plains to the east are divided longitudinally by the Alleghanny Mountains, which stretch from the Carolinas to the Gulf of St. Lawrence, parallel to the Atlantic, and at no great distance from it. Although the general direction of the two chains is from south to north, yet, as they maintain a degree of parallelism to the two coasts, they diverge towards the north, one inclining towards the north-west, and the other towards the north-east. The long narrow plain between the Atlantic and the Alleghannies is divided, throughout its length, by a line of cliffs not more than 200 or 300 feet above the Atlantic plain—the outcropping edge of the Second Terrace, or Atlantic Slope, whose rolling surface goes west to the foot of the mountains.

An enormous table-land occupies the greater part of Mexico, or Anahuac. It begins at the Isthmus of Tehuantepec, and extends north-west to the 42nd parallel of north latitude, a distance of about 1600 miles, which is nearly equal to the distance from the north extremity of Scotland to Gibraltar. It is narrow towards the south, but expands towards the north-west till about the latitude of the city of Mexico, where it attains its greatest breadth of 360 miles, and there also it is highest. The most easterly part in that parallel is 7500 feet above the sea, from whence it rises towards the west to the height of 9000 feet at the city of Mexico, and then gradually diminishes to 4000 feet towards the Pacific.

Its height in California is not known, but it still bears the character of a table-land, and maintains an elevation of 6000 feet along the east side of the Sierra Madre, even to the 32nd degree of north latitude, where it sinks to a lower level before joining the Rocky Mountains. The descent from this plateau to the low lands is very steep on all sides; on the east, especially, it is so precipitous that, from a distance, it is like a range of high mountains. There are only two carriage-roads to it from the Mexican Gulf, by passes 500 miles asunder: one at Xalapa, near Vera Cruz; the other at Santilla, west of Monterey. The descent to the shores of the Pacific is almost equally rapid, and that to the south no less so, where, for 300 miles between the plains of Tehuantepec and the Rio Yapez, it presses on the shores of the Pacific, and terminates in high mountains, leaving only a narrow margin of hilly maritime coast. Where the surface of the table-land is not traversed by mountains it is as level as the ocean. There is a carriage-road over it for 1500 miles, without hills, from the city of Mexico to Santa Fé.

The southern part of the plateau is divided into four parts, or distinct plains, surrounded by hills from 500 to 1000 feet high. In one of these, the plain of Tolesco, on a small group of islands near the eastern bank of the Lake Tetzcuco, and surrounded by a wall of porphyritic mountains, stands the city of Mexico, once the capital of the empire of Montezuma, which must have far surpassed the modern city in extent and splendour, as many remains of its ancient glory testify. It is 9000 feet above the sea, which is the height of Mount St. Bernard.

One of the singular crevices through which the internal fire finds a vent stretches from the Gulf of Mexico to the Pacific, directly across the table-land, in a line about 16 miles south of the city of Mexico. A very remarkable row of active volcanos occurs along this parallel. Turtla, the most eastern of them, is in the 95th degree west longitude, near the Mexican Gulf, in a low range of wooded hills. More to the west the snow-shrouded cone of Orizabo is 17,000 feet high; and its ever-fiery crater, seen like a star in the darkness of the night, has obtained it the name of Citlaltepetel, the "Mountain of the Star." Popocatepetl, the loftiest mountain in Mexico, 17,884 feet above the sea, lies still farther west, and is in a state of constant eruption. A chain of smaller volcanos unites the three. On the western slope of the table-land, 36 leagues from the Pacific, stands the volcanic cone of Jorullo, on a plain 2890 feet above the sea. It suddenly appeared and rose 1683 feet above the plain on the night of the 29th of September, 1759. The great cone of Colima, the last of this volcanic series, stands insulated in the plain of that name, between the western declivity of the table-land and the Pacific.

A high range of mountains extends along the eastern margin of the table-land to Real de Catorce, and the surface of the high plain is divided into two parts by the Sierra Madre, which begins at 21 degrees north latitude; and, after going north about 60 miles, its continuity is broken into the insulated ridges of the Sierra Altamina, and the group containing the mines of Zacatecas; it soon after resumes its character of a regular chain, and, with a breadth of 100 miles, proceeds in parallel ridges and longitudinal valleys to New Mexico, where it skirts both banks of the Rio Bravo del Norte, and joins the Sierra Verde, the most southern part of the Rocky Mountains, in 40 degrees north latitude.

To the south some points of the Sierra Madre are said to be 10,000 feet high, and 4000 above their base; and between the parallels of 36 and 42 degrees, where the chain is the watershed between the Rio Colorado and the Rio Bravo del Norte, they are still higher, and perpetually covered with snow. The mountains on the left bank of the last-mentioned river are the eastern ridges of the Sierra Madre, and contain the sources of the innumerable affluents of the Missouri and other rivers that flow into the Mississippi and Mexican Gulf.

Deep cavities, called Barancas, are a characteristic feature of the table-lands of Mexico. They are long narrow rents two or three miles in breadth, and many more in length, often descending 1000 feet below the surface of the plain, with a brook or the tributary of some river flowing through them. Their sides are precipitous and rugged, with overhanging rocks covered with large trees. The intense heat adds to the contrast between these hollows and the bare plains, where the air is more than cool.

Vegetation varies with the elevation; consequently the splendour which adorns the low lands vanishes on the high plains, which, though producing much grain and pasture, are often saline, sterile, and treeless, except in some places, where oaks grow to an enormous size free of underwood.

The Rocky Mountains run 1500 miles, in two parallel chains, from the Sierra Verde to the mouth of the Mackenzie River, in the Arctic Ocean, sometimes united by a transverse ridge. In some places the eastern range rises to the snow-line, and even far above it, as in Mounts Hooper and Brown, 15,590 and 16,000 feet above the sea; but the general elevation is only above the line of trees. The western range is not so high till north of the 55th parallel, where both ranges are of the same height, and frequently higher than the snow-line. They are generally barren, though the transverse valleys have fertile spots with grass, and sometimes trees. The long valley between the two rows of the Rocky Mountains, which is 100 miles wide, must have considerable elevation in the south, since the tributaries of the Colombia River descend from it in a series of rapids and cataracts for nearly 100 miles; and it is probably still higher towards the sources of the Peace River, where the mountains, only 1500 feet above it, are perpetually covered with snow. The Sierra Verde is 670 miles from the Pacific; but, as the coast trends due north to the Sound of Juan de Fuca, the western range of the Rocky Mountains maintains a distance of 380 miles from the ocean, from that point to the latitude of Behring's Bay in 60 degrees north latitude.

Offsets from the Sierra Madre, and the volcanic group of Castres Virgines, fill the peninsula of California, from whence, to the Sound of Juan de Fuca, the Pacific is bordered by snow-clad mountains. Prairies extend between this coast-chain and the Rocky Mountains from California to north of the Oregon River. The Oregon coast for 200 miles is a mass of undisturbed forest-thickets and marshes, and north from it, with few exceptions, is a mountainous region of bold aspect, often reaching above the snow-line. The maritime chain of Russian

America, of a still more Alpine character, runs due north to 60 degrees of north latitude, where Mount Elias rises to 17,000. The branch which runs westward to Bristol Bay has many active volcanos, and so has that which fills the promontory of Alaska.

The archipelagos and islands along the coast, from California to the promontory of Alaska, have the same bold character as the mainland, and may be regarded as the tops of a submarine chain of table-lands and mountains, which constitute the most westerly ridge of the maritime chains. Prince of Wales's Archipelago contains seven active volcanos.

The mountains on the coasts of the Pacific, and the islands are, in many places, covered with colossal forests, but wide tracts in the south are sandy deserts.

CHAPTER XII.

NORTH AMERICA CONTINUED—THE GREAT CENTRAL PLAINS OR VALLEY OF THE MISSISSIPPI — THE ALLEGHANNY MOUNTAINS—THE ATLANTIC SLOPE — THE ATLANTIC PLAIN—GEOLOGICAL NOTICE.

THE great central plain of North America, lying between the Rocky and Alleghanny Mountains, and reaching from the Gulf of Mexico to the Arctic Ocean, includes the valleys of the Mississippi, St. Lawrence, Nelson, Churchill, and most of those of the Missouri, Mackenzie's, and Coppermine rivers. It has an area of 3,240,000 square miles, which is 240,000 square miles more than the central plain of South America, and about half the size of the great plain of the Old Continent, which is less fertile; for, although the whole of America is not more than half the size of the Old Continent, it contains at least as much productive soil.

This plain, 5000 miles long, becomes wider towards the north, and has no elevations, except a low table-land which crosses it at the line of the Canadian lakes and the sources of the Mississippi, and is nowhere above 1500 feet high, and rarely more than 700. The character of the plain is that of perfect uniformity, rising by a gentle regular ascent from the Gulf of Mexico to the sources of the Mississippi, which river is the great feature of the North

American low lands. The ground rises in the same equable manner from the right bank of the Mississippi to the foot of the Rocky Mountains, but its ascent from the left bank to the Alleghannies is broken into hill and dale, containing the most fertile territory in the United States. Under so wide a range of latitude the plain embraces a great variety of soil, climate, and productions; but, being almost in a state of nature, it is characterized in its middle and southern parts by interminable grassy savannahs, or prairies, and enormous forests; and in the far north by deserts which rival those of Siberia in dreariness.

In the south a sandy desert, 400 or 500 miles wide, stretches along the base of the Rocky Mountains to the 41st degree N. lat. The dry plains of Texas and the upper region of the Arkansas have all the characteristics of Asiatic table-lands; more to the north the bare, treeless steppes on the high grounds of the far west are burnt up in summer, and frozen in winter by biting blasts from the Rocky Mountains; but the soil improves towards the Mississippi. At its mouth, indeed, there are marshes which cover 35,000 square miles, bearing a rank vegetation, and its delta is a labyrinth of streams and lakes, with dense brushwood. There are also large tracts of forest and saline ground, but all the cultivation on the right bank of the river is along the Gulf of Mexico and in the adjacent provinces, and is entirely tropical, consisting of sugar-cane, cotton, and indigo. The prairies, so characteristic of North America, then begin.

To the left of the Mississippi these savannahs are sometimes rolling, but oftener level and interminable as the ocean, covered with long rank grass of tender green, blended with flowers chiefly of the liliaceous kind, which fill the air with their fragrance. In the southern districts they are sometimes interspersed with groups of magnolia, tulip and cotton-trees; and in the north, oaks and black walnut. These are rare occurrences, as the prairies may be traversed for many days without finding a shrub, except on the banks of the streams, which are beautifully fringed with myrtle, azalea, kalmea, andromeda, and rhododendron. On the wide plains the only objects to be seen are countless herds of wild horses, buffalos and deer. The country assumes a more severe aspect in higher latitudes. It is still capable of producing rye and barley in the territories of the Assinniboines, and round Lake Winnepeg there are great forests; a low vegetation, with grass, follows, and towards the Icy Ocean the land is barren and covered with numerous lakes.

East of the Mississippi there is a magnificent undulating country about 300 miles broad, extending 1000 miles from south to north between that great river and the Alleghanny Mountains, mostly covered with trees. When America was discovered, one uninterrupted forest spread over the country, from the Gulf of St. Lawrence and the Canadian lakes to the Gulf of Mexico, and from the Atlantic Ocean it crossed the Alleghanny Mountains, descended into the valley of the Mississippi on the north, but in the south it crossed the main stream of that river altogether, forming an ocean of vegetation of more than 1,000,000 square miles, of which the greater part still remains. Although forests occupy so much of the country, there are immense prairies on the east side of the river also. Pine-barrens, stretching far into the interior, occupy the whole coast of the Mexican Gulf eastward from the Pearl River, through Alabama and a great part of Florida.

These vast monotonous tracts of sand, covered with forests of gigantic pine-trees, are as peculiarly a distinctive feature of the continent of North America as the prairies, and are not confined to this part of the United States; they occur to a great extent in North Carolina, Virginia, and elsewhere. Tennessee and Kentucky, though much cleared, still possess large forests, and the Ohio flows for hundreds of miles among magnificent trees, with an undergrowth of azaleas, rhododendrons, and other beautiful shrubs, matted together by creeping plants. There the American forests appear in all their glory, the gigantic deciduous cypress, and the tall tulip-tree, overtopping the forest by half its height, a variety of noble oaks, black walnuts, American plàne, hiccory, sugar-maple, and the lyriodendron, the most splendid of the magnolia tribe, the pride of the forest.

The Illinois waters a country of prairies ever fresh and green, and five new states are rising round the great lakes, whose territory of 280,000 square miles contains 180,000,000 acres of land, of excellent quality. These states, still mostly covered with wood, lie between the lakes and the Ohio, and they reach from the United States to the Upper Mississippi—a country twice as large as France, and six times the size of England.

The quantity of water, in the north-eastern part of the central plain, greatly preponderates over that of the land; the five principal lakes, Huron, Superior, Michigan, Erie, and Ontario, cover an area equal to Great Britain, without reckoning small lakes and rivers innumerable.

The Canadas contain millions of acres of good soil, covered with immense forests. Upper Canada is the most fertile, and in many respects is one of the most valuable of the British colonies in the west: every European grain, and every plant that requires a hot summer, and can endure a cold winter, thrives there. The forests consist chiefly of black and white spruce, the Weymouth and other pines—trees which do not admit of undergrowth: they grow to great height, like bare spars, with a tufted crown, casting a deep gloom below. The fall of large trees from age is a common occurrence, and not without danger, as it often causes the destruction of those adjacent, and an ice-storm is awful.

After a heavy fall of snow, succeeded by rain and a partial thaw, a strong frost coats the trees and all their branches with transparent ice often an inch thick: the noblest trees bend under the load, icicles hang from every bough, which come down in showers with the least breath of wind. The hemlock-spruce especially, with its long drooping branches, is then like a solid mass. If the wind freshens, the smaller trees become like corn beaten down by the tempest, while the large ones swing heavily in the breeze. The forest at last gives way under its load; tree comes down after tree with sudden and terrific violence, crushing all before them till the whole is one wide uproar, heard from afar like successive discharges of artillery. Nothing, however, can be imagined more brilliant and beautiful than the effect of sunshine in a calm day on the frozen boughs, where every particle of the icy crystals sparkles, and nature seems decked in diamonds.*

Although the subsoil is perpetually frozen at the depth of a few feet below the surface, beyond the 56th degree of North latitude, yet trees grow in some places up to the 64th parallel. Farther north, the gloomy and majestic forests cease, and are succeeded by a bleak, barren waste, which becomes progressively more dreary as it approaches the Arctic Ocean. Four-fifths of it are like the wilds of Siberia in surface and climate, covered many months in the year with deep snow. During the summer it is the resort of herds of rein-deer and buffalos, which come from the south to browse on the tender short grass which then springs up along the streams and lakes.

The Alleghanny or Appalachian chain, which constitutes the second or subordinate system of North American mountains, separates the great central

* Mr. Taylor.

plain from that which lies along the Atlantic Ocean. Its base is a strip of table-land from 1000 to 3000 feet high, lying between the sources of the rivers Alabama and Yazan, in the southern states of the Union, and New Brunswick, at the mouth of the river St. Lawrence. This high land is traversed throughout 1000 miles, between Alabama and Vermont, by from three to five parallel ridges of low mountains rarely more than 3000 or 4000 feet high, and separated by fertile longitudinal valleys, which occupy more than two-thirds of its breadth of 100 miles. In Virginia and Pennsylvania, the only part of the chain to which the name of the Alleghanny Mountains properly belongs, it is 150 miles broad; and the whole is computed to have an area of 2,000,000 square miles. The parallelism of the ridges, and the uniform level of their summits, are the characteristics of this chain, which is lower and less wild than the Rocky Mountains. The uniformity of outline in the southern and middle parts of the chain is very remarkable, and results from their peculiar structure.* These mountains have no central axis, but consist of a series of convex and concave flexures, forming alternate hills and longitudinal valleys, running nearly parallel throughout their length, and cut transversely by the rivers that flow to the Atlantic on one hand, and to the Mississippi on the other. The water-shed nearly follows the windings of the coast, from the point of Florida to the north-western extremity of the State of Maine.

* Mr. Lyell's America.

The picturesque and peaceful scenery of the Appalachian Mountains is well known; they are generally clothed with a luxuriant and varied vegetation, and their western slope is considered one of the finest countries in the United States. To the south they maintain a distance of 200 miles from the Atlantic, but approach close to the coast in the south-eastern part of the state of New York, from whence their general course is northerly to the river St. Lawrence. They fill the Canadas, Maine, New Brunswick, and Nova Scotia with branches as high as the mean elevation of the principal chain, and extend even to the dreary regions of Baffin's Bay. Not only the deep forests, but vegetation in general, diminish as the latitude increases, till on the Arctic shores the soil becomes incapable of culture, and the majestic forest is superseded by the Arctic birch which creeps on the ground. The islands along the north-eastern coasts have more than the mildness of the main-land. Though little favoured by nature, many of them produce flax and timber; and Newfoundland, as large as England and Wales, maintains a population of 70,000 souls by its fisheries; it is nearer to Britain than any part of America—the distance from the port of St. John to the harbour of Valentia in Ireland is only 1656 nautical miles.

The long, and comparatively narrow plain which lies between the Appalachian Mountains and the Atlantic, extends from the Gulf of Mexico to the eastern coast of Massachusetts. At its southern extremity it joins the plain of the Mississippi, and gradually becomes narrower in its northern course to New England, where it merely includes the coast islands. It is divided throughout its length by a line of cliffs from 200 to 300 feet high, which begins in Alabama, and ends in the coast of Massachusetts. This escarpment is the eastern edge of the terrace known as the Atlantic Slope, which rises above the Maritime or Atlantic Plain, and undulates westward to the foot of the Blue Mountains, the most eastern ridge of the Appalachian Chain. It is narrow at its extremities in Alabama and New York, but in Virginia and the Carolinas it is 200 miles wide. The surface of the slope is of great uniformity; ridges of hills and long valleys run along it parallel to the mountains, close to which it is 600 feet high. It is rich in soil and cultivation, and has an immense water-power in the streams and rivers flowing from the mountains across it, which are precipitated over its rocky edge to the plains on the west. More than twenty-three rivers of considerable size fall in cascades down this ledge between New York and the Mississippi, affording scenes of great beauty.

Both land and water assume a new aspect on the Atlantic Plain. The rivers, after dashing over the rocky barrier, run in tranquil streams to the ocean, and the plain itself is a monotonous level, not more than a hundred feet above the surface of the sea. Along the coast it is scooped into valleys and ravines, with innumerable creeks.

The greater part of the magnificent countries east

of the Alleghannies is in a high state of cultivation and commercial prosperity, with natural advantages not surpassed in any country. Nature, however, still maintains her sway in some parts, especially where pine-barrens and swamps prevail. The territory of the United States occupies 7,000,000 or 8,000,000 square miles, the greater part of it capable of producing everything that is useful to man, but not more than the twenty-sixth part of it has been cleared; the climate is healthy, the soil fertile, abounding in mineral treasures, and it possesses every advantage from navigable rivers and excellent harbours. The outposts of civilization have already advanced half-way to the Pacific, and the tide of white men is continually and irresistibly pressing onwards to the ultimate extinction of the original proprietors of the soil—a melancholy, but not a solitary instance of the rapid extinction of a whole race.

Crystalline and silurian rocks, rich in precious and other metals, form the substratum of Mexico, for the most part deeply covered with plutonic and volcanic formations and secondary limestone; yet granite comes to the surface on the coast of Acapulca, and occasionally on the plains and mountains of the table-land. The Rocky Mountains are mostly silurian, except the eastern ridge, which is of stratified crystalline rocks, amygdaloid and ancient volcanic productions. The coast-chain has the same character, with immense tracts of volcanic rocks, both ancient and modern, especially obsidian, which is nowhere developed on a greater scale, except in Mexico and the Andes.

In North America, as in the southern part of the continent, volcanic action is entirely confined to the coast and highland along the Pacific. The numerous vents in Mexico and California are often in great activity, and hot springs abound. Though a considerable interval occurs north of these, where the fire is dormant, the country is full of igneous productions, and it again finds vent in Prince of Wales's Island, which has seven active volcanos. From Mount St. Elias westward through the whole southern coast of the peninsula of Russian America and the Aleutian Islands which form a semicircle between Cape Alaska, in America, and the peninsula of Kamtchatka, volcanic vents occur, and in the latter peninsula there are three of great height.

From the similar nature of the coasts, and the identity of the fossil mammalia on each side of Behring's Strait, it is more than probable that the two continents were united even since the sea was inhabited by the existing species of shell-fish. Some of the gigantic quadrupeds of the Old Continent are supposed to have crossed either over the land or over the ice to America, and to have wandered southward through the longitudinal valleys of the Rocky Mountains, Mexico, and Central America, and to have spread over the vast plains of both continents, even to their utmost extremity. An extinct species of horse, the mastodon, a species of elephant, three gigantic edentata, and a hollow-horned ruminating animal roamed over the pampas of the southern continent, and the prairies of the northern; certainly since the sea was peopled by its present inhabitants, probably even since the existence of the Indians. The skeletons of these creatures are found in great numbers in the saline marshes on the prairies called the Licks, which are still the resort of the existing races.

There were, however, various animals peculiar to America, as well as to each part of that continent, at least as far as yet known. South America still retains in many cases the type of its ancient inhabitants, though on a very reduced scale. But on the Patagonian plains and on the pampas, skeletons of creatures of gigantic size and anomalous forms have been found; one like an ant-eater of great magnitude, covered with a prodigious coat of mail similar to that of the armadillo; others like gigantic rats or mice, perhaps the largest animals yet discovered,—all of which had lived on vegetables, and had existed at the same time with those already mentioned. These animals were not destroyed by the agency of man, since creatures not larger than a rat vanished from Brazil within the same period.

The geological outline of the United States, the Canadas, and all the country to the Polar Ocean, though highly interesting in itself, becomes infinitely more so when viewed in connection with that of northern and middle Europe. A remarkable analogy exists in the structure of the land on each side of the north Atlantic basin. Gneiss, mica-schist, and occasional granite, prevail over wide areas in the Alleghannies, on the Atlantic slope, and still more in the northern latitudes of the American continent; and they range also through the greater part of Scandinavia, Finland, and Lapland. In the latter countries, and in the more northern parts of America, Mr. Lyell has observed that the fossiliferous rocks belong either to the most ancient or to the newest formations, to the Silurian strata, or to such as contain shells of recent species only, no intermediate formation appearing through immense regions. Silurian strata extend over 2000 miles in the middle and high latitudes of North America; they occupy a tract nearly as great between the most westerly headlands of Norway and those that separate the White Sea from the Polar Ocean; and Sir Roderick Murchison has traced them through central and eastern Europe, and the Ural Mountains, even to Siberia. Throughout these vast regions, both in America and Europe, the Silurian strata are followed in ascending order by the Devonian and carboniferous formations, which are developed on a stupendous scale in the United States, chiefly in the Alleghanny Mountains and on the Atlantic slope. The Devonian and carboniferous strata together are a mile and a half thick in New York, and three times as much in Pennsylvania, where one single coal-field occupies 63,000 square miles between the northern limits of that State and Alabama. There are many others of great magnitude, both in the States and to the north of them, so that the most valuable of all minerals is here inexhaustible, which is not the least of the many advantages enjoyed by that flourishing country. The coal formation is also developed in New Brunswick, and traces of it are found on the shores and in the islands of the Polar Ocean, on the east coast of Greenland, and even in Spitzbergen.

Vast carboniferous basins exist in Belgium above the Silurian strata; and a great portion of Britain is perfectly similar in structure to North America. The Silurian rocks in many instances are the same; and the coal-fields of New England are precisely similar to those in Wales, 3000 miles off. It would be difficult to estimate the quantity of coal in Britain and Ireland, but there is probably enough to last for some thousand years. If science continues to advance as it has lately done, a substitute will in all probability be discovered before the coal is exhausted.

In all the more northern countries that have been mentioned, so very distant from one another, the general range of the rocks is from north-east to south-west; and in northern Europe, the British isles, and North America, great lakes are formed along the junction of the strata, the whole analogy affording a proof of the wide diffusion of the same geological conditions in the northern regions at a very remote period. At a later time those erratic blocks, which are now scattered over the higher latitudes of both continents, were most likely brought from the north by drift ice or currents, while the land was still covered by the deep. Volcanic agency has not been wanting to complete the analogy. The Silurian and overlying strata have been pierced in many places by trappean rocks in both continents, and they appear also in the islands of the North Atlantic and Polar seas. Even now the volcanic fires are in great activity in the very centre of that basin in Iceland, and in the very distant and less known island of Jan Mayen's Land.

CHAPTER XIII.

GREENLAND — SPITZBERGEN — ICELAND—JAN MAYEN'S LAND —ANTARCTIC LANDS—VICTORIA CONTINENT.

GREENLAND, the most extensive of the Arctic lands, begins with the lofty promontory of Cape Farewell, the southern extremity of a group of rocky islands, which are separated by a channel five miles wide from a table-land of appalling aspect, narrow to the south, but increasing in breadth northward to a distance of which only 1300 miles are known. This table-land is bounded by mountains rising from the deep in mural precipices, which terminate in needles and pyramids, or in parallel terraces of alternate snow and bare rock, occasionally leaving a narrow shore. The coating of ice is so continuous and thick that the surface of the table-land may be regarded as one enormous glacier, which overlaps the rocky edges and dips between the mountain peaks into the sea.

The coasts are beset with rocky islands, and cloven by fiords which, in some instances, wind like rivers for 100 miles into the interior. These deep inlets of the sea, now sparkling in sunshine, now shaded in gloom, are hemmed in by walls of rock often 2000 feet high, whose summits are hid in the clouds. They generally terminate in glaciers, which are sometimes forced on by the pressure of the upper ice plains till they fill the fiord and even project far into the sea like bold headlands, when, undermined by the surge, huge masses of ice fall from them with a crash like thunder, making the sea boil. These icebergs, carried by currents, are stranded on the Arctic coast, or are driven into lower latitudes. The ice is very transparent and compact in the Arctic regions: its prevailing tints are blue, green, and orange, which, contrasted with the dazzling whiteness of the snow and the gloomy hue of the rocks, produce a striking effect.

A great fiord in the 68th parallel of latitude is supposed to extend completely across the table-land, dividing the country into south and north Greenland, which last extends indefinitely towards the pole, but it is altogether inaccessible from the frozen sea and the iron-bound shore, so that, excepting a very small portion of the coast, it is an unknown region.

In some sheltered spots in south Greenland, especially along the borders of the fiords, there are meadows where the service-tree bears fruit; beech and willow trees grow by the streams, but not taller than a man; and still farther north the willow and juniper scarcely rise above the surface; yet this country has a flora peculiar to itself. South of the island of Disco, on the west coast, Danish colonies and missionaries have made settlements on some of the islands, and at the mouths of fiords; the Esquimaux inhabit the coasts even to the extremity of Baffin's Bay.

The aspect of other Arctic lands is like that of Greenland. In the island of Spitzbergen the mountains spring sharp and grand from the margin of the sea in dark gloomy masses, mixed with pure snow and enormous glaciers, presenting a sublime spectacle. The sun is not seen for several months in the year, when the intensity of the cold splits rocks and makes the sea reek like a boiling caldron. Many have perished in the attempt to winter in this island, yet a colony of Russian hunters and fishermen lead a miserable existence there within 10° of the pole—the most northern inhabited spot on the globe.

Although the direct rays of the sun are powerful in sheltered spots within the Arctic circle, the thermometer does not rise above 45° of Fahrenheit. July is the only month in which snow does not fall, and in the end of August the sea at night is covered with a thin coating of ice, and a summer often passes without one day that can be called warm. The snow-blink, the aurora, the stars, and the moon, which appears ten or twelve days without intermission in her northern declination, furnish the greatest light the inhabitants enjoy in their long winter.

Iceland is 200 miles east from Greenland, and lies south of the Arctic Circle, which its most northern point touches. Though a fifth part larger than Ireland, not more than 4000 square miles are habitable; all beside being a chaos of volcanos and ice.*

* Trevelyan's Travels in Iceland.

The peculiar feature of Iceland lies in a trachytic region, which seems to rest on an ocean of fire. It consists of two vast parallel table-lands covered with ice-clad mountains, stretching from N.E. to S.W. through the very centre of the island, separated by a longitudinal valley nearly 100 miles wide, which reaches from sea to sea. These mountains assume rounded forms with long level summits, or domes with sloping declivities, as in the trachyte mountains of the Andes and elsewhere; but such huge masses of tufa and conglomerate project from their sides in perpendicular or overhanging precipices, separated by deep ravines, that the regularity of their structure can only be perceived from a distance: they conceal under a cold and tranquil coating of ice the fiery germs of terrific convulsions, sometimes bursting into dreadful activity, sometimes quiescent for ages. The most extensive of the two parallel ranges of Jokuls or Ice Mountains runs along the eastern side of the valley, and contains Ordefa, the highest point in Iceland, seen like a white cloud from a great distance at sea; the western high land passes through the centre of the island.

Glaciers cover many thousand square miles in Iceland, descending from the mountains and pushing far into the low lands. This tendency of the ice to encroach has very materially diminished the quantity of habitable ground, and the progress of the glaciers is facilitated by the influence of the ocean of subterranean fire, which heats the superincumbent ground and loosens the ice.

The longitudinal space between the mountainous table-lands is a low valley 100 miles wide, extending from sea to sea, where a substratum of trachyte is covered with lava, sand, and ashes, studded with low volcanic cones. It is a tremendous desert, never approached without dread even by the natives; a scene of perpetual conflict between the antagonist powers of fire and frost, without a drop of water or a blade of grass: no living creature is to be seen, not a bird nor even an insect. The surface is a confused mass of streams of lava rent by crevices; and rocks piled on rocks, with occasional glaciers, complete the scene of desolation. As herds of rein-deer are seen browsing on the Iceland moss that grows plentifully at its edges, it may be presumed that some unknown parts may be less barren. The extremities of the valley are more especially the theatres of perpetual volcanic activity. At the southern end, which opens to the sea in a wide plain, there are many volcanos, of which Hekla is most known, from its insulated position, its vicinity to the coast, and its tremendous eruptions. The cone is divided into three peaks by crevices which are filled with snow: one of these fissures cleaves the mountain from the summit to the base; it is supposed to have been produced by the great eruption of 1300. Between the years 1004 and 1766 twenty-three violent eruptions have taken place, one of which continued six years, spreading devastation over a country once the abode of a thriving colony, now covered with lava, scoriæ, and ashes; and in the year 1846 it was in full activity. The eruption of Skaptar, which broke out on the 8th of May, 1783, and continued till August, is one of the most dreadful recorded. The sun was hid many days by dense clouds of vapour, which extended to England and Holland, and the quantity of matter thrown out in this eruption was computed at fifty or sixty thousand millions of cubic yards. Some rivers were heated to ebullition, others dried up: the condensed vapour fell in snow and torrents of rain; the country was laid waste, famine and disease ensued, and in the course of the two succeeding years 1300 people and 150,000 sheep and horses perished. The scene of horror was closed by a dreadful earthquake. Previous to the explosion an ominous mildness of temperature indicated the approach of the volcanic fire towards the surface of the earth: similar warnings had been observed before in the eruptions of Hekla.

A semicircle of volcanic mountains, on the eastern side of the lake Myvatr, is the focus of the igneous phenomena at the northern end of the great central valley. Leirhnukr and Krabla, on the N.E. of the lake, have been especially formidable. After years of quiescence they suddenly burst into violent eruption, and poured such a quantity of lava into the lake Myvatr, which is 20 miles in circumference, that the water boiled many days. There are other volcanos in this district no less formidable. Various caldrons of boiling mineral pitch, the shattered craters of ancient volcanos, occur at the base of this semicircle of mountains, and also on the flanks of Mount Krabla. These caldrons throw up jets of the dark matter, enveloped in clouds of steam, at regular intervals, with a loud explosion.

The eruptive boiling springs of Iceland are perhaps the most extraordinary phenomenon in this singular country. All the great aqueous eruptions occur in the trachytic formation: they are characterized by their high temperature, by holding siliceous matter in solution, which they deposit in the form of siliceous sinter, and by the discharge of sulphuretted hydrogen gas. Numerous instances of spouting springs occur at the extremities of the great central valley, especially at its southern end where more than fifty have been counted in the space of a few acres—some constant, others periodical, some merely agitated, or stagnant. The Great Geyser and Stokr, six miles north-west from Hekla, are the most magnificent; at regular intervals they project large columns of boiling water 100 feet high, enveloped in clouds of steam, with tremendous noise. Some springs emit gas only, or gas with a small quantity of water. Such fountains are not confined to the land, or fields of ice; they occur also in the sea, and many issue from crevices in the lava-bed of the lake Myvatr, and rise in jets above the surface of the water.

A region of the same character with the mountains of the Icelandic desert extends due west from it to the extremity of the long narrow promontory of the Sneefield Syssel, ending in the snow-clad cone of the

Sneefield Jokul, 5000 feet high, one of the most conspicuous mountains in Iceland.

With the exception of the purely volcanic districts described, trap-rocks cover 20,000 square miles of Iceland, in beds perfectly parallel, and almost horizontal, which have been formed by streams of lava at very ancient epochs, spread over the country occasionally 4000 feet deep.

The dismal coasts are torn in every direction by fiords penetrating many miles into the interior, and splitting into endless branches. In these fissures the sea is still, dark, and deep between walls of rock 1000 feet high. The fiords, however, do not here, as in Greenland, terminate in glaciers, but are prolonged in narrow valleys through which streams and rivers run to the sea. In these valleys the inhabitants have their abode, or in meadows which have a transient verdure along some of the fiords, where the sea is so deep that ships find safe anchorage.

In the valleys on the northern coast, near as they approach to the Arctic circle, the soil is wonderfully good, and there is more vegetation than in any other part of Iceland, with the exception of the eastern shore, which is the most favoured portion of this desolate land. Rivers abounding in fish are much more frequent there than elsewhere; willows and juniper adorn the valleys, and birch-trees 20 feet high grow in the vale of Lagerflest, the only place which produces them large enough for house-building, and the verdure is fine on the banks of those streams which are heated by volcanic fires.

The climate of Iceland is much less rigorous than that of Greenland, and it would be still milder were not the air chilled by the immense fields of ice from the Polar Sea which beset its shores.

The inhabitants are supplied with fuel by the Gulf Stream, which brings drift wood in great quantity from Mexico, the Carolinas, Virginia, the river St. Lawrence, some even from the Pacific Ocean, is drifted by currents round by the northern shores of Siberia. The mean temperature in the south of the island is about 39° of Fahrenheit, that of the central districts 36°, and in the north it is rarely above the freezing point. The cold is most intense when the sky is clear, but that is a rare occurrence, as the wind from the sea covers mountain and valley with thick fog. Hurricanes are frequent and furious, and, although thunder is seldom heard in high latitudes, Iceland is an exception, for tremendous thunder-storms are not uncommon there—a circumstance no doubt owing to the volcanic nature of that island, as lightning accompanies volcanic eruptions everywhere. The sun is always above the horizon in the middle of summer, and under it in midwinter, yet there is no absolute darkness.

The island of Jan Mayen lies nearly midway between Iceland and Spitzbergen: it is the most northern volcanic country known. Its principal feature is the volcano of Beerenberg, 6870 feet high, flanked by enormous glaciers, whose lofty snow-capped cone, apparently inaccessible, has been seen to emit fire and smoke.

The south polar lands are equally volcanic, and as deeply icebound, as those to the north. Victoria Land, which from its extent seems to form part of a continent, was discovered by Sir James Ross, who commanded the expedition sent by the British Government in 1839 to ascertain the position of the south magnetic pole. The extensive tract lies under the meridian of New Zealand; Cape North, its most northern point, is situate in 70° 31′ S. lat., and 165° 28′ E. long. To the west of that cape the northern coast of this new land terminates in perpendicular ice-cliffs from 200 to 500 feet high, strerching as far as the eye can reach, with a chain of grounded icebergs extending for miles from the base of the cliffs, all of tabular form, and varying in size from one to nine or ten miles in circumference. A lofty range of peaked mountains rises in the interior at Cape North, covered with unbroken snow, only relieved from uniform whiteness by shadows produced by the undulations of the surface. The indentations of the coast are filled with ice many hundreds of feet thick, which makes it impossible to land. To the east of Cape North the coast trends first to S.E. by E., and then in a southerly direction to $78\frac{1}{4}$° of south latitude, at which point it suddenly bends to the east and extends in one contiuuous vertical ice-cliff to an unknown distance in that direction. The first view of Victoria Land is described as most magnificent. "On the 11th of January, 1841, in about latitude 71° S., and longitude 171° E., the Antarctic con-

tinent was first seen, the general outline of which at once indicated its volcanic character, rising steeply from the ocean in a stupendous mountain range, peak above peak, enveloped in perpetual snow, and clustered together in countless groups resembling a vast mass of crystallisation, which, as the sun's rays were reflected on it, exhibited a scene of such unequalled magnificence and splendour as would baffle all power of language to portray or give the faintest conception of. One very remarkable peak, in shape like a huge crystal of quartz, rose to the height of 7867 feet, another to 9096, and a third to 8444 feet above the level of the sea. From these peaks ridges descended to the coast, terminating abruptly in bold capes and promontories, whose steep escarpments, affording shelter to neither ice nor snow, alone showed the jet black lava or basalt which reposed beneath the mantle of eternal frost." . . . .
"On the 28th, in latitude 77° 31′, and longitude 167° 1′, the burning volcano, Mount Erebus, was discovered covered with ice and snow from its base to its summit, from which a dense column of black smoke towered high above the numerous other lofty cones and crateriferous peaks with which this extraordinary land is studded from the 72nd to the 78th degree of latitude. Its height above the sea is 12,367 feet; and Mount Terror, an extinct crater adjoining it, which has doubtless once given vent to fires beneath, attains an altitude little inferior, being 10,884 feet in height, and ending in a cape from which a vast barrier of ice extended in an easterly

direction, checking all farther progress south. This continuous perpendicular wall of ice, varying in height from 200 to 100 feet, its summit presenting an almost unvarying level outline, we traced for 300 miles, when the pack-ice obstructed all farther progress." *

The vertical cliff in question forms a completely solid mass of ice about 1000 feet thick: the greater part of which is below the surface of the sea; there is not the smallest appearance of a fissure throughout its whole extent; and the intensely blue sky beyond, indicated plainly the great distance to which the ice-plains reach southwards. Gigantic icicles hang from every projecting point of the icy cliff, showing that it sometimes thaws in these latitudes, although in the month of February, which corresponds with August in England, Fahrenheit's thermometer did not rise above 14° at noon. In the North Polar Ocean, on the contrary, streams of water flow from every iceberg during summer. The whole of this country is beyond the pale of vegetation: no moss, not even a lichen, covers the barren soil, where everlasting winter reigns. Parry's Mountains, a lofty range stretching south from Mount Terror to the 79th parallel, is the most southerly land yet discovered. The south magnetic pole, the object of the expedition, is situated in Victoria Land, in 75° 5′ S. lat., and 154° 8′ E. long.

* Remarks on the Antarctic Continent and Southern Islands, by Robert M'Cormick, Esq., Surgeon of H. M. S. Erebus.

Various tracts of land have been discovered near the Antarctic circle, and within it, though none in so high a latitude as Victoria Land: whether they form part of one large continent remains to be ascertained. Discovery ships, which have been sent by the Russian, French, and American Governments, have increased our knowledge of these far regions, and the spirited adventures of British merchants and captains of whalers have contributed quite as much.

The land within the Arctic circle is generally volcanic, at least the coast-line, which is all that is yet known, and, being covered with snow and ice, it is destitute of vegetation.

CHAPTER XIV.

THE CONTINENT OF AUSTRALIA—TASMANIA, OR VAN DIEMEN'S LAND—NEW ZEALAND—NEW GUINEA—BORNEO—ATOLLS—ENCIRCLING REEFS—CORAL REEFS—BARRIER REEFS—VOLCANIC ISLANDS—AREAS OF SUBSIDENCE AND ELEVATION IN THE BED OF THE PACIFIC—ACTIVE VOLCANOS.

The labyrinth of islands that is scattered over the Pacific Ocean for more than 30 degrees on each side of the equator, and from the 130th eastern meridian to Sumatra, which all but unites this enormous archipelago to the continent of Asia, has the group of New Zealand or Tasmania, and the continent of Australia, with its appendage, Van Diemen's Land, on the south; and altogether forms a region which, from the unstable nature of the surface of the earth, is partly the wreck of a continent that has been engulfed by the ocean, and partly the highest summits of a new one rising above the waves. This extensive portion of the globe is, in many parts, terra incognita; the Indian Archipelago has never been explored, and, with the exception of our colonies in New Holland and New Zealand, is little known.

The continent of New Holland, 2400 miles from east to west, and 1700 from north to south, is divided into two unequal parts by the tropic of Capricorn, and consequently has both a temperate and a tropical climate. New Guinea, separated from New Holland by Torres Straits, and traversed by the same chain of mountains with New Holland and Van Diemen's Land, is so perfectly similar in structure, that it forms but a detached member of the adjacent continent.

The coasts of New Holland are indented by very large bays, and by harbours that might give shelter to all the navies in Europe. The most distinguishing feature of the eastern side, which is chiefly occupied by the British colony of New South Wales, is a long chain of mountains which never goes far from the coast, and, with the exception of some short deviations in its southern part, maintains a meridional direction through 35° of latitude. It is continued at one extremity from Torres Straits, at the north end of the Gulf of Carpentaria, far into the interior of New Guinea; and at the other it traverses the whole of Van Diemen's Land. It is low in the northern parts of New Holland, being in some places merely a high land; but about the 30th degree of south latitude it assumes the form of a regular mountain chain, and, running in a very tortuous line from N.E. to S.W., terminates its visible course at Wilson's Promontory, the southern extremity of the continent. It is continued, however, by a chain of mountainous islands across Bass's Straits to Cape Portland, in Van Diemen's Land; and from thence the range proceeds in a zigzag line of high and picturesque mountains to South Cape, where it ends, having, in its course of 1500 miles, separated the drainage of both countries into eastern and western waters.

The distance of the chain from the sea in New South Wales is from 50 to 100 miles, but at the 32nd parallel it recedes to 150, yet soon returns, and forms the wild group of the Corecudgy peaks, from whence, under the names of the Blue Mountains and Australian Alps, its highest part, it proceeds in a general westerly direction to the land's ends.

The average height of these mountains is only from 2400 to 4700 feet above the level of the sea, and even Mount Kosciuszko, the loftiest of the Australian Alps, is not more than 6500 feet high, yet its position is so favourable, that the view from its snowy and craggy top sweeps over 7000 square miles. The rugged and savage character of these mountains far exceeds what might be expected from their height: in some places, it is true, their tops are rounded and covered with forests; but by far the greater part of the chain, though wooded along the flanks, is crowned by naked needles, tooth-formed peaks, and flat crests of granite or porphyry, mingled with patches of snow. The spurs give a terrific character to these mountains, and in many places render them altogether inaccessible, both in New South Wales and Van Diemen's Land. These shoot right and left from the ridgy axis of the main range, equal to it in height, and separated from it, and from one another, by dark and almost subterraneous gullies, like rents in the bosom of the earth, iron-bound by impracticable precipices, with streams flowing through them in black silent eddies or foaming torrents. The intricate character of these ravines, the danger of descending into them, and the difficulty of getting out again, render this mountain-chain, in New South Wales at least, almost a complete barrier between the country on the coast and that in the interior—a circumstance very unfavourable to the latter.*

In New South Wales the country slopes westward from these mountains to a low, flat, unbroken plain. On the east side, darkly verdant and round-topped hills and ridges are promiscuously grouped together, leading to a richly-wooded undulating country, which gradually descends to the coast, and forms the valuable lands of the British colony. Discovered by Cook in the year 1770, it was not colonized till 1788. It has become a prosperous country; and although new settlers in the more remote parts suffer the privations and difficulties incident to their position, yet there is educated society in the towns, with the comforts and luxuries of civilized life.

The coast-belt on the western side of New Holland is generally of inferior land, with richer tracts interspersed near the rivers; and bounded on the east by a range of primary mountains from 3000 to 4000 feet high, in which granite occasionally appears. Beyond this the country is level, and the land better, though nowhere very productive except in grass.

* Memoirs of Count Strzelecki.

None of the rivers of New Holland are navigable to any great distance from their mouths; the want of water is severely felt in the interior, which, as far as it is known, is a treeless desert of sand, swamps, and jungle; yet a belief prevails that there is a large sea, or fresh-water lake, in its centre; and this opinion is founded partly on the nature of the soil, and also because all the rivers that flow into the sea on the northern coast, between the gulfs of Van Diemen and Carpentaria, converge towards their sources, as if they served for drains to some large body of water.

However unpropitious the middle of the continent may be, and the shores generally have the same barren character, there is abundance of fine country inland from the coasts. On the north all tropical productions might be raised, and in so large a continent there must be extensive tracts of arable land, though its peculiar character is pastoral. There are large forests on the mountains and elsewhere, yet that moisture is wanting which clothes other countries in the same latitudes with rank vegetation. In the colonies the clearing of a great extent of land has increased the mean annual temperature, so that the climate has become hotter and drier, and not thereby improved.

Van Diemen's Land, of triangular form, has an area of 27,200 square miles, and is very mountainous. No country has a greater number of deep commodious harbours; and as most of the rivers, though not navigable to any distance, end in arms of the sea, they afford secure anchorage for ships of any size. The mountain-chain that traverses the colony of New South Wales, and the islands in Bass's Straits, starts anew from Cape Portland, and, winding through Van Diemen's Land in the form of the letter Z, separates it into two nearly equal parts, with a mean height of 3750 feet, and at an average distance of 40 miles from the sea. It encloses the basins of the Derwent and Heron rivers, and, after sending a branch between them to Hobart Town, ends at South Cape. The offsets which shoot in all directions are as savage and full of impassable chasms as it is itself. There are cultivable plains and valleys along the numerous rivers and large lakes by which the country is well watered; so that Van Diemen's Land is more agricultural and fertile than the adjacent continent, but its climate is wet and cold. The uncleared soil of both countries, however, is far inferior to that in the greater part of North or South America.*

Granite constitutes the entire floor of the western portion of New South Wales, and extends far into the interior of the continent, bearing a striking resemblance in character to a similar portion of the Altaï chain described by Baron Humboldt. The central axis of the mountain range, in New South Wales and in Van Diemen's Land, is of granite, syenite, and quartz; but in early times there had been great invasions of volcanic substances, as many

* Count Strzelecki.

parts of the main chain, and most of its offsets, are of the older igneous rocks. The fossiliferous strata of the two colonies are mostly of the Palæozoic period, but their fossil fauna is poor in species. Some are identical with, and others are representatives of, the species of other countries, even of England. It appears, from their coal-measures, that the flora of these countries was as distinct in appearance from that of the northern hemisphere, previous to the carboniferous period, as it is at the present day.

New Zealand, divided into three islands by rocky and dangerous channels, is superior to Australia in richness of soil, fertility, and beauty, and abounds in fine timber and a variety of vegetable and mineral productions. High mountains run through the islands, which in the most northerly rise 14,000 feet above the stormy ocean around, buried two-thirds of their height in permanent snow and glaciers, and exhibiting on the grandest scale all the Alpine characters, with the addition of active volcanos on the eastern and western coasts. The coast is a broken country, overspread with a most luxuriant, but dark and gloomy vegetation. There are undulating tracts and table-lands of great extent without a tree, overrun by ferns and a low kind of myrtle; but the mountain-ridges are clothed with dense and gigantic forests. There is much good land and many lakes, with navigable rivers and the best of harbours; so that this country is peculiarly well suited for a colony, but difficult of access from a boisterous ocean.

A very different scene from the stormy seas of New Zealand presents itself to the north of Australia. There, vivified by the glowing sun of the equator, the islands of the Indian Archipelago are of matchless beauty, crowned by lofty mountains, loaded with aromatic verdure, that shelve to the shore, or dip into a transparent glassy sea. Their coasts are cut by deep inlets, and watered by the purest streams, which descend in cascades, rushing through wild crevices. The whole is so densely covered with palms and other beautiful forms of tropical vegetation, that they seem to realize a terrestrial paradise.

Papua, or New Guinea, is the largest island in the Pacific, 1400 miles long, and 200 in width, with mountains rising above mountains, till in the west they attain the height of 16,000 feet, capped with snow, and two volcanos burn on its northern shores. From its position so near the equator, it is probable that New Guinea has the same vegetation with the Spice Islands to the east; and, from the little that is known of it, must be one of the finest countries in existence.

Borneo, next in size to New Guinea, is a noble island, divided in two by the equator, and traversed through its whole length by magnificent chains of mountains, which end in three branches at the Java Sea. Beautiful rivers flow from them to the plains, and several of these spring from a spacious lake on the table-land in the interior, among the peaks of Keni-Balu, the highest point of the island.

Diamonds, gold, and antimony are among its minerals; gums, precious woods, and all kinds of spices and tropical fruits, are among its vegetables.

A volume might be written on the beauty and riches of the Indian Archipelago. Many of the islands are hardly known, and the interior of the greater number has never been explored; so that they offer a wide field of discovery to the enterprising traveller, and they are now of easier access since the seas have been cleared of pirates by the Honorable Captain Keppel. The success of Mr. Brook in conciliating the natives is a noble instance of the power of mind.

They have become of much importance since our relation with China has been altered, and on that account Captain Stanley, and other scientific naval officers, have been employed to survey the coasts and channels of these unknown seas. The great intertropical islands in the Pacific, likewise other large islands, as Ceylon and Madagascar in the Indian Seas, which, by the way, do not differ in character from the preceding, are really continents in miniature, with their mountains and plains, their lakes and rivers; and in climate they vary, like the main land, with the latitude, only that continental climates are more extreme both as to heat and cold.

It is a singular circumstance, arising from the instability of the crust of the earth, that, with only three or four exceptions, all the smaller tropical islands in the Pacific and Indian Oceans are either volcanic or coralline, except New Caledonia and the Seychelles; and it is a startling fact, that, in most cases where there are volcanos, the land is rising by slow and almost imperceptible degrees above the ocean, whereas there is every reason to believe that those vast spaces, studded with coral islands or atolls, are actually sinking below it, and have been for ages.*

There are four different kinds of coral formations in the Pacific and Indian Oceans, all entirely produced by the growth of organic beings and their detritus; namely, lagoon islands or atolls, encircling reefs, barrier reefs, and coral fringes. They are all nearly confined to the tropical regions; the atolls to the Pacific and Indian Oceans alone.

An atoll, or lagoon island, consists of a chaplet or ring of coral, enclosing a lagoon, or portion of the ocean, in its centre. The average breadth of the part of the ring above the surface of the sea is about a quarter of a mile, oftener less, and it seldom rises higher than from 6 to 10 or 12 feet above the waves. Hence the lagoon islands are not discernible at a very small distance, unless when they are covered with the cocoa-nut, palm, or the pandana, which is frequently the case. On the outer side this ring or circlet shelves down to the distance of 100 or 200 yards from its edge, so that the sea gradually deepens to 25 fathoms, beyond which the sides plunge at once into the unfathomable depths of the ocean, with a more rapid descent than the cone

* Darwin on Coral Reefs.

of any volcano. Even at the small distance of some hundred yards, no bottom has been found with a sounding-line a mile and a half long. All the coral at a moderate depth below water is alive—all above is dead, being the detritus of the living part washed up by the surf, which is so tremendous on the windward side of the tropical islands of the Pacific and Indian Oceans that it is often heard miles off, and is frequently the first warning to seamen of their approach to an atoll.

On the lagoon side, where the water is calm, the bounding-ring, or reef, shelves into it by a succession of ledges, also of living coral, though not of the same species with those which build the exterior wall and the foundations of the whole ring. The perpetual change of water brought into contact with the external coral by the breakers probably supplies them with more food than they could obtain in a quieter sea, which may account for their more luxuriant growth. At the same time, they deprive the whole of the corals in the interior of the most nourishing part of their food, because the still-water in the lagoon, being supplied from the exterior by openings in the ring, ceases to produce the hardier corals; and species of more delicate forms, and of much slower growth, take their place.* The depth of the lagoon varies, in different atolls, from 20 to 50 fathoms, the bottom being partly detritus and

* Supplement to the Observations on the Temple of Serapis, by Charles Babbage, Esq.

partly live coral. By the growth of the coral, some few of the lagoons have been filled up; but the process is very slow from the causes assigned, and also because there are marine animals that feed on the living coral, and prevent its indefinite growth. In all departments of nature, the exuberant increase of any one class is checked and limited by others. The coral is of the most varied and delicate structure, and of the most beautiful tints. Dark brown, vivid green, rich purple, pink, deep blue, peach-colour, yellow, with dazzling white, contrasted with deep shadows, shine through the limpid water; while fish of the most gorgeous hues swim among the branching coral, which are of many different kinds, though all combine in the structure of these singular islands. Lagoon islands are sometimes circular, but more frequently oval or irregular in their form. Sometimes they are solitary, or in groups, but they occur most frequently in elongated archipelagos, with the atolls elongated in the same direction. The grouping of atolls bears a perfect analogy to the grouping of the archipelagos of ordinary islands.

The size of atolls varies from two to ninety miles in diameter, and islets are frequently formed on the coral rings by the washing up of the detritus, for they are so low that the waves break over them in high tides or storms. They have openings or channels in their circuit, generally on the lee side, where the tide enters, and by these ships may sail into the lagoons, which are excellent harbours; and even on the surface of the circlet or reef itself there are occasionally boat-channels between the islets.

Dangerous Archipelago, lying east of the Society Islands, is one of the most remarkable assemblages of atolls in the Pacific Ocean. There are 80 of them, generally of a circular form, surrounding very deep lagoons, and separated from each other by profound depths. The reefs or rings are about half a mile wide, and seldom rise more than 10 feet above the edge of the surf, which beats on them with such violence that it may be heard at the distance of 8 miles; and yet on that side the coral insects build more vigorously, and vegetation thrives better, than on the other: many of the islets are inhabited.

The Caroline Archipelago, the largest of all, lies north of the equator, and extends its atolls in 60 groups over 1000 miles. Many are of great size, and all are beat by a tempestuous sea and occasional hurricanes. The atolls in the Pacific Ocean and China Sea are beyond enumeration. Though less frequent in the Indian Ocean, none are more interesting, or afford more perfect specimens of this peculiar formation, than the Maldiva and Laccadive Archipelagos, both nearly parallel to the coast of Malabar, and elongated in that direction. The former is 470 miles long, and about 50 miles broad, with the atolls arranged in a double row, separated by an unfathomable sea, into which their sides descend with more than ordinary rapidity. The largest atoll is 88 miles long, and somewhat less than 20 broad; Suadiva, the next in size, is 44 miles by 23, with a large lagoon in its centre, to which there is access by 42 openings. There are inhabited islets on most of the chaplets or rings not higher than 20 feet, while the reefs themselves are nowhere more than 6 feet above the surge.

The Laccadives run to the north of this archipelago in a double line of nearly circular atolls, on which are low inhabited islets.

Encircling reefs differ in no respect from atoll reefs except that they have one or more islands in their lagoon. They commonly form a ring round mountainous islands, at a distance of two or three miles from the shore, rising on the outside from a very deep ocean, and separated from the land by a lagoon or channel 200 or 300 feet deep. These reefs surround the submarine base of the island, and, rising by a steep ascent to the surface, they encircle the island itself. The Caroline Archipelago, already mentioned, exhibits good examples of this structure in the encircled islands of Hogolen and Seniavine: the narrow ring or encircling reef of the former is 135 miles in its very irregular circuit, on which are a vast number of islets; but six or eight islands rise to considerable height from its lagoon, which is so deep, and the opening into it so large, that a frigate might sail into it. The encircling reef of Seniavine is narrow and irregular, and its lagoon is so nearly filled by a lofty island, that it leaves only a strip of water round it from two to five miles wide and 30 fathoms deep.

Otaheite, the largest of the Society group, is another instance of an encircled island of the most beautiful kind; it rises in mountains 7000 feet high, with only a narrow plain along the shore, and, except where cleared for cultivation, it is covered with forests of cocoa-nut, palms, bananas, bread-fruit, and other productions of a tropical climate. The lagoon, which encompasses it like an enormous moat, is 30 fathoms deep, and is hemmed in from the ocean by a coral band of the usual kind, at a distance varying from half a mile to three miles.

Barrier reefs are of precisely the same structure as the two preceding classes, from which they only differ in their position with regard to the land. A barrier reef off the north-east coast of the continent of Australia is the grandest coral formation existing. Rising at once from an unfathomable ocean, it extends 1000 miles along the coast, with a breadth varying from 200 yards to a mile, and at an average distance of from 20 to 30 miles from the shore, in some places increasing to 60 and even 70 miles. The great arm of the sea included between it and the land is nowhere less than 10, occasionally 60 fathoms deep, and is safely navigable throughout its whole length, with a few transverse openings, by which ships can enter. The reef is really 1200 miles long, because it stretches nearly across Torres Straits. There are also extensive barrier reefs on the islands of Louisiade and New Caledonia, which are exactly opposite to the great Australian reef; and as atolls stud that part of the Pacific which lies between them, it is called the Coralline Sea. The rolling of the billows along the great Australian reef has been admirably described. "The long ocean-swell, being suddenly impeded by this barrier, lifted itself in one great continuous ridge of deep blue water, which, curling over, fell on the edge of the reef in an unbroken cataract of dazzling white foam. Each line of breaker runs often one or two miles in length with not a perceptible gap in its continuity. There was a simple grandeur and display of power and beauty in this scene that rose even to sublimity. The unbroken roar of the surf, with its regular pulsation of thunder, as each succeeding swell fell first on the outer edge of the reef, was almost deafening, yet so deep-toned as not to interfere with the slightest nearer and sharper sound. . . . . . Both the sound and sight were such as to impress the spectator with the consciousness of standing in the presence of an overwhelming majesty and power."*

Coral reefs are distinct from all the foregoing: they are merely fringes of coral along the margin of a shore, and, as they line the shore itself, they have no lagoons. A vast extent of coast, both on the continents and islands, are fringed by these reefs, and, as they frequently surround shoals, they are very dangerous.

Lagoon islands are the work of various species

* By Mr. Jukes, Naturalist to the Surveying Voyage of Captain Blackwood, R.N., in Torres Straits.

of coral insects, but those particular polypi which build the profound external wall, the foundation and support of the whole ring or reef, are most vigorous when most exposed to the breakers: they cannot exist at a greater depth than 25 or 30 fathoms at most, and die immediately when left dry; yet the coral wall descends precipitously to unfathomable depths; and although the whole of it is not the work of these insects, yet the perpendicular thickness of the coral is known to be very great, extending hundreds of feet below the depth at which these polypi cease to live. From an extensive survey of the Coralline Seas of the tropics, Mr. Darwin has found an explanation of these singular phenomena in the instability of the crust of the earth.

Since there are certain proofs that large areas of the dry land are gradually rising, and others sinking down, so the bottom of the ocean is not exempt from the general change that is slowly bringing about a new state of things; and as there is evidence on multitudes of the volcanic islands in the Pacific of a rise in certain parts of the basis of the ocean, so the lagoon islands indicate a subsidence in others—changes arising from the expansion and contraction of the strata under the bed of the ocean.

There are strong reasons for believing that a continent once occupied a great part of the tropical Pacific, some part of which subsided by slow and imperceptible degrees. As portions of it gradually sank down below the surface of the deep, the tops of mountains and table-lands would remain as islands of different magnitude and elevation, and would form archipelagos elongated in the direction of the mountain-chains. Now the coral-insect which constructs the outward wall and mass of the reefs never builds laterally, and cannot exist at a greater depth than 25 or 30 fathoms. Hence, if it began to lay the foundations of its reef on the submerged flanks of an island, it would be obliged to build its wall upwards in proportion as the island sank down, so that at length a lagoon would be formed between it and the land. As the subsidence continued, the lagoon would increase, the island would diminish, and the base of the coral reef would sink deeper and deeper, while the insects would always keep its top just below the surface of the ocean, till at length the island would entirely disappear, and a perfect atoll would be left. If the island were mountainous, each peak would form a separate island in the lagoon, and the encircled islands would have different forms, which the reefs would follow continuously. This theory perfectly explains the appearances of the lagoon islands and barrier reefs, the continuity of the reef, the islands in the middle of the lagoons, the different distances of the reefs from them, and the forms of the archipelago so exactly similar to the archipelagos of ordinary islands, all of which are but the tops of submerged mountain-chains, and generally partake of their elongated forms.

Every intermediate form between an atoll and an encircling reef exists; New Caledonia is a link between them. A reef runs along the north-western coast of that island 400 miles, and for many leagues never approaches within 8 miles of its shore, and the distance increases to 16 miles near the southern extremity. At the other end the reefs are continued on each side 150 miles beyond the submarine prolongation of the land marking the former extent of the island. In the lagoon of Keeling Atoll, situate in the Indian Ocean 600 miles south of Sumatra, many fallen trees and a ruined store-house show that it has subsided: these movements take place during the earthquakes at Sumatra, which are also felt in this atoll. Violent earthquakes have lately been felt at Vanikora, a lofty island with an encircling reef in the western part of the South Pacific, and on which there are marks of recent subsidence. Other proofs are not wanting of this great movement in the beds of the Pacific and Indian Oceans.

The extent of the atoll formations, including under this name encircling reefs, is enormous. In the Pacific, from the southern end of Low Archipelago to the northern end of Marshall Archipelago, a distance of 4500 miles, and many degrees of latitude in breadth, there is not an island that is not of atoll formation. The same may be said of the space in the Indian Ocean between Saya de Matha and the end of the Laccadives, which includes 25 degrees of latitude—such are the enormous areas that have been, and probably still are, slowly subsiding. Other spaces of great extent may also be mentioned—as the large archipelago of the Carolinas, that in the Coralline Sea off the north-west coast of Australia, and an extensive one in the China Sea.

Though the volcanic islands in the Pacific are so numerous, there is not one within the areas mentioned, and there is not an active volcano within several hundred miles of an archipelago, or even group of atolls. This is the more interesting, as recent shells and fringes of dead coral, found at various heights on their surfaces, show that the volcanic islands have been rising more and more above the surface of the ocean for a very long time.

The volcanic islands also occupy particular zones in the Pacific, and it is found from extensive observation that all the points of eruption fall on the areas of elevation.

One of the most terribly active of these zones begins with the Banda group of islands, and includes Timor, Sumbawa, Bali, Java, and Sumatra, separated only by narrow channels, and altogether forming a gently curved line 2000 miles long; but as the volcanic zone is continued through Barren Island, in the Bay of Bengal, northward to an island off the Birmah coast, the entire length of this volcanic range is a great deal more.

The little island of Gounong-Api, belonging to the Banda group, contains a volcano of great activity; and such is the elevating pressure of the submarine fire in that part of the ocean, that a mass of black basalt rose up of such magnitude as to fill a bay 60 fathoms deep so quietly that the inhabitants were not aware of what was going on till it was nearly done. Timor and the other adjacent islands also bear marks of recent elevation.

There is not a spot of its size on the face of the earth that contains so many volcanos as the island of Java.* A range of volcanic mountains, from 5000 to 13,000 feet high, forms the central crest of the island, and ends to the east in a series of 38 separate volcanos with broad bases rising gradually into cones. They all stand on a plain but little elevated above the sea, and each individual mountain seems to have been formed independently of the rest. Most of them are of great antiquity, and are covered with thick vegetation. Some are extinct, or only emit smoke; from others sulphureous vapours issue with prodigious violence; one has a large crater filled with boiling water; and a few have had fierce eruptions of late years. The island is covered with volcanic spurs from the main ridge, united by cross chains, together with other chains of less magnitude but no less fury.

In 1772 the greater part of one of the largest volcanic mountains was swallowed up after a short but severe combustion: a luminous cloud enveloped the mountain on the 11th of August, and soon after the huge mass actually disappeared under the earth with tremendous noise, carrying with it about 90 square miles of the surrounding country, 40 villages, and 2957 of their inhabitants.

* Sir Stamford Raffles on Java.

The northern coast of Java is flat and swampy, but the southern provinces are beautiful and romantic; yet in the lovely peaceful valleys the stillness of night is disturbed by the deep roaring of the volcanos, many of which are perpetually burning with slow but terrific action.

Separated by narrow channels of the sea, Bali and Sumbawa are but a continuation of Java, the same in nature and structure, but on a smaller scale, their mountains being little more than 8000 feet high.

The intensity of the volcanic force under this part of the Pacific may be imagined from the eruption of Tomboro in Sumbawa in 1815, which continued from the 5th of April till July: the explosions were heard at the distance of 970 miles; and in Java, at the distance of 300 miles, the darkness during the day was like that of deep midnight. The country around was ruined, and the town of Tomboro was submerged by heavy rollers from the ocean.

In Sumatra the extensive granitic formations of eastern Asia join the volcanic series which occupies so large a portion of the Pacific. This most beautiful of islands presents the boldest aspect: it is indented by arms of the most transparent sea, and watered by innumerable streams; it displays in its vegetation all the bright colouring of the tropics. Here the submarine fire finds vent in three volcanos on the southern, and one on the northern side of the island. A few atolls, many hundreds of miles to the south, show that this volcanic zone alternates with an area of subsidence.

More to the north, and nearly parallel to the preceding zone, another line of volcanic islands begins to the north of New Guinea, and passes through New Britain, New Ireland, Solomon's Island, and the New Hebrides, containing many open vents. This range, or area of elevation, separates the Coralline Sea from the great chain of atolls on the north between Ellice's group and the Caroline Islands, so that it lies between two areas of subsidence.

The third and greatest of all the zones of volcanic islands begins at the northern extremity of Celebes, and includes Gilolo, one of the Molucco group, which is bristled with volcanic cones; and from thence it may be traced northwards through the Philippine Islands and Formosa: bending thence to the north-east, it passes through Loo Choo, the Japan Archipelago, and is continued by the Kurile Islands to the peninsula of Kamtchatka, where there are several active volcanos of great elevation.

The Philippine Islands and Formosa form the volcanic separation between the atoll region in the China Sea and that of the Caroline and Pellew groups.

There are six islands east of Jephoon, in the Japan Archipelago, which are subject to eruptions, and the internal fire breaks through the Kurile Islands in 18 vents, besides having raised two new islands in the beginning of this century, one four miles round and the other 3000 feet high, though the ocean there is so deep that the bottom has not been reached with a line 200 fathoms long.

Thus some long rent in the earth had reached from the tropics to the gelid seas of Okhotsk, probably connected with the peninsula of Kamtschatka: a new one begins to the east of the latter in the Aleutian Islands, which are of the most barren and desolate aspect, perpetually beaten by the surge of a restless ocean, and bristled by the cones of 24 volcanos; they sweep in a half-moon round Behring's Sea till they join the volcanic peninsula of Russian America.

The line of volcanic agency has been followed far beyond the limits of the coral working insects, which extend but a short way on each side of the tropics; but it has been shown that, in the equatorial regions, immense areas of elevation alternate with as great areas of subsidence: north of New Holland they are so mixed that it indicates a point of convergence.*

On the other side of the Pacific the whole chain of the Andes, and the adjacent islands of Juan Fernandez and the Galapagos, form a vast volcanic area, which is actually now rising. And though there are few volcanic islands north of the zone of atolls, yet those that be indicate great internal activity, especially the Sandwich Islands, where the volcanos of Owhyhee are inferior to none in awful sublimity.

* Darwin on Volcanic Islands.

It may be observed that, where there are coral fringes, the land is either rising or stationary; for, were it subsiding, lagoons would be formed. On the contrary, there are many fringing reefs on the shores of volcanic islands along the coasts of the Red Sea, the Persian Gulf, and the West Indian islands, all of which are rising. Indeed, this occurrence, in numberless instances, coincides with the existence of upraised organic remains on the land.

As the only coral formations in the Atlantic are fringing reefs, the bed of that ocean is not sinking; and, with the exception of the Leeward Islands, the Canaries, and Cape de Verde groups, there are no active volcanos on the islands or on the coasts of that ocean. The Peak of Teneriffe is a splendid instance.

At present the great continent has few centres of volcanic action in comparison with what it once had. The Mediterranean is still undermined by fire, which occasionally finds vent in Vesuvius and the stately cone of Etna. Though Stromboli constantly pours forth an inexhaustible stream of lava, and a temporary island now and then starts up from the sea, the volcanic action is diminished, and Italy has become comparatively more tranquil.

The table-land of western Asia, especially Azerbijan, had once been the seat of intense commotion, now spent, or only smoking from the snowy cone of Demavend. The table-land of eastern Asia furnishes the solitary instance of igneous explosion at a distance from the sea in the volcanic chain of the Thean-Tchan.

The seat of activity has been perpetually changing. There always has been volcanic action, possibly more intense in former times, but even at present it extends from pole to pole.

Notwithstanding the numerous volcanic vents in the globe, many places are subject to violent earthquakes, which ruin the works of man, and often change the configuration of the country.

Earthquakes are produced by fractures and sudden heavings and subsidences in the elastic crust of the globe, from the pressure of the liquid fire, vapour, and gases in its interior, which there find vent, relieve the tension which the strata acquire during their slow refrigeration, and restore equilibrium. But whether the initial impulse be eruptive, or a sudden pressure upwards, the shock originating in that point is propagated through the elastic surface of the earth in a series of circular or oval undulations, similar to those produced by dropping a stone into a pool, and like them they become broader and lower as the distance increases, till they gradually subside: in this manner the shock travels through the land, becoming weaker and weaker till it terminates. When the impulse begins in the interior of a continent, the elastic wave is propagated through the solid crust of the earth, as well as in sound through the air, and is transmitted from the former to the ocean, where it is finally spent and lost, or, if very powerful, is continued in the opposite land. Almost all the great earthquakes however have their origin in the bed of the ocean, far from land, whence the shocks travel in undulations to the surrounding shores.

No doubt many of small intensity are imperceptible; it is only the violent efforts of the internal forces that can overcome the pressure of the ocean's bed, and that of the superincumbent water. The internal pressure is supposed to find relief most readily in a belt of great breadth that surrounds the land at a considerable distance from the coast, and, being formed of its débris, the internal temperature is in a perpetual state of fluctuation, which would seem to give rise to sudden flexures and submarine eruptions.

When the original impulse is a fracture or eruption of lava in the bed of the deep ocean, two kinds of waves or undulations are produced and propagated simultaneously—one through the bed of the ocean, which is the true earthquake shock: and coincident with this a wave is formed and propagated on the surface of the ocean, which rolls to the shore, and reaches it in time to complete the destruction long after the shock or wave through the solid ocean-bed has arrived and spent itself on the land. The height to which the surface of the ground is elevated, or the vertical height of the shock-wave, varies from one inch to two or three feet. This earth-wave, on passing under deep water, is imperceptible, but when it comes to soundings it carries with it to the land a long flat aqueous wave: on arriving at the beach the water drops in arrear from the superior velocity of the shock, so that at that moment the sea seems to recede before the great ocean-wave arrives.

It is the small forced wave that gives the shock to ships, and not the great wave; but when ships are struck in very deep water, the centre of disturbance is either immediately under, or very nearly under, the vessel.

Three other series of undulations are formed simultaneously with the preceding, by which the sound of the explosion is conveyed through the earth, the ocean, and the air, with different velocities. That through the earth travels at the rate of from 7000 to 10,000 feet in a second in hard rock, and somewhat less in looser materials, and arrives at the coast a short time before, or at the same moment with the shock, and produces the hollow sounds that are the harbingers of ruin; then follows a continuous succession of sounds, like the rolling of distant thunder, formed, first, by the wave that is propagated through the water of the sea, which travels at the rate of 4700 feet in a second; and, lastly, by that passing through the air, which only takes place when the origin of the earthquake is a submarine explosion, and travels with a velocity of 1123 feet in a second. The rolling sounds precede the arrival of the great wave on the coasts, and are continued after the terrific catastrophe when the eruption is extensive.

When there is a succession of shocks all the phenomena are repeated.

The velocity of the great oceanic wave varies as the square root of the depth ; it consequently has a rapid progress through deep water, and less when it comes to soundings. The velocity of the shock varies with the elesticity of the strata it passes through. The undulations of the earth are subject to the same laws as those of light and sound ; hence, when the shock or earth-wave passes through strata of different elasticity, it will partly be reflected, and a wave will be sent back, producing a shock in a contrary direction, and partly refracted, or its course changed ; so that shocks will occur both upwards and downwards, to the right or to the left of the original line of transit. Hence most damage is done at the junction of deep alluvial plains with the hard strata of the mountains, as in the great earthquake in Calabria in the year 1783.

When the height of the undulations is small, the earthquake will be a horizontal motion, which is the least destructive ; when the height is great, the vertical and horizontal motions are combined, and the effect is terrible ; but the worst of all is a verticose or twisting motion, which nothing can resist. It is occasioned by the crossing of two waves of horizontal vibration, which unite at their point of intersection and form a rotatory movement. This, and the interferences of shocks arriving at the same point from different origins or routes of different length, account for the repose in some places, and those extraordinary phenomena that took place during the earthquake of 1783 in Calabria, where the shock diverged on all sides from a centre through a highly elastic base covered with alluvial soil, which was tossed about in every direction. The dynamics of earthquakes are ably discussed by Mr. Mallet in a very interesting paper in the Transactions of the Royal Irish Academy.

There are few places where the earth is long at rest; for, independently of those secular elevations and subsidences that are in progress over such extensive tracts of country, small earthquake shocks must be much more frequent than we imagine, though imperceptible to our senses, and only to be detected by means of instruments. The shock of an earthquake at Lyons in February, 1822, was not generally perceptible at Paris, yet the wave reached and passed under that city, and was detected by the swinging of the large declination needle at the Observatory, which had previously been at rest. Even in Scotland 139 slight shocks have been registered within a few years, of which 81 occurred at Comrie in Perthshire, but the cause is at no great depth under the surface, as the shocks extended to a small distance.

The undulations of some of the great earthquakes have spread to an enormous extent: that which destroyed Lisbon had its origin immediately under the devoted city, from whence the shock extended over an area of about 700,000 square miles, or a twelfth part of the circumference of the globe: the West Indian islands, and the lakes in Scotland, Norway, and Sweden, were agitated by it. It began without warning, and in five minutes the city was a heap of ruins.

The earthquake of 1783, in Calabria, which completely changed the face of the country, lasted only two minutes, but it was not very extensive. Baron Humboldt's works are full of interesting details on this subject, especially with regard to the tremendous convulsions in South America.

Sometimes a shock has been carried underground which was not felt at the surface, as in the year 1802, in the silver-mine of Marienberg, in the Hartz. In some instances miners have been insensible to shocks felt on the surface above, which happened at Fahlun, in Sweden, in 1823—circumstances depending, in both instances, on the elasticity of the strata, the depth of the impulses, or obstacles that may have changed the course of the terrestrial undulation. During earthquakes dislocations of strata take place, the course of rivers is changed, and in some instances they have been permanently dried up, rocks are hurled down, masses raised up, and the configuration of the country altered; but if there be no fracture at the point of original impulse, there will be no noise.

CHAPTER XV.

THE OCEAN—ITS SIZE, COLOUR, PRESSURE, AND SALTNESS—TIDES, WAVES, AND CURRENTS—TEMPERATURE—NORTH AND SOUTH POLAR ICE—INLAND SEAS.

THE ocean, which fills a deep cavity in the globe and covers three-fourths of its surface, is so unequally distributed that there is three times more land in the northern than in the southern hemisphere. The torrid zone is chiefly occupied by sea, and only one twenty-seventh part of the land on one side of the earth has land opposite to it on the other. The form assumed by this immense mass of water is that of a spheroid flattened at the poles; and as its mean level is always nearly the same, for anything we know to the contrary, it serves as a base for measuring the height of the land.

The bed of the ocean, like that of the land, of which it is the continuation, is diversified by plains and mountains, table-lands and valleys, sometimes barren, sometimes covered with marine vegetation, and teeming with life. Now it sinks into depths which the sounding-line has never fathomed, now it appears in chains of islands, or rises near to the surface in hidden reefs and shoals, perilous to the mariner. Springs of fresh water rise from the bottom, volcanos eject their lavas and scoriæ, and earthquakes trouble the deep waters.

The ocean is continually receiving the spoils of the land, and from that cause would constantly be decreasing in depth, and, as the quantity of water is always the same, its superficial extent would increase: there are however counteracting causes to check this tendency; the secular elevation of the land over extensive tracts, in many parts of the world, is one of the most important.* Volcanos, coral islands, and barrier reefs show that great changes of level are constantly taking place in the bed of the ocean itself, —that symmetrical bands of subsidence and elevation extend alternately over an area equal to a hemisphere, from which it may be concluded that the balance is always maintained between the sea and land, although the distribution may vary in the lapse of time.

The Pacific or Great Ocean exceeds in superficies all the dry land on the globe. It has an area of 50,000,000 square miles: including the Indian Ocean its area is nearly 70,000,000. Its breadth from Peru to the coast of Africa is 16,000 miles: it is shorter than the Atlantic, as it only communicates with the Arctic Ocean by Behring's Strait, whereas the Atlantic, as far as we know, stretches from pole to pole.

The continent of Australia occupies a comparatively small portion of the Pacific, while innumerable islands stud its surface many degrees on either side of the equator, of which a great number are

* Darwin on Coral Reefs.

volcanic, showing that its bed has been, and indeed actually is, the theatre of violent igneous eruptions. So great is its depth that a line five miles long has not reached the bottom in many places. Between the tropics it is generally unfathomable; yet, as the whole mass of the ocean counts for little in the total amount of terrestrial gravitation, its mean depth is but a small fraction of the radius of the globe.

The bed of the Atlantic is a long deep valley with few mountains, or at least but few that raise their summits in islands above its surface. Its greatest breadth, including the Gulf of Mexico, is 5000 miles, and its superficial extent is about 25,000,000 square miles. This sea is exceedingly deep. In 27° 26′ S. lat. and 17° 29′ W. long. Sir James Ross found the depth to be 14,550 feet; 450 miles west from the Cape of Good Hope it was 16,062 feet, or 332 feet more than the height of Mont Blanc; and in 15° 3′ S. lat. and 23° 14′ W. long. a line of 27,600 feet did not reach the bottom, which is equal to the height of some of the most elevated peaks of the Himalaya, but there is reason to believe that many parts of the ocean are still deeper. A great part of the German Ocean is only 93 feet deep, though on the Norwegian side, where the coast is bold, the depth is 910 fathoms.

Immense sand-banks often project from the land, which rise from great depths to within a few fathoms of the surface. Of these the Aghullus Bank, at the Cape of Good Hope, is one of the most remarkable: that off Newfoundland is still greater; it consists of a double bank, which is supposed to reach to the north of Scotland. The Dogger Bank, in the North Sea, and many others, are well known: some on the coast of Norway are surrounded by such deep water that they must be submarine table-lands. All are the resort of fish.

The pressure at great depths is enormous. In the Arctic Ocean, where the specific gravity of the water is least, on account of the melting of the ice, the pressure at the depth of a mile and a quarter is 2809 pounds on a square inch of surface: this was confirmed by Captain Scoresby, who says, in his 'Arctic Voyages,' that the wood of a boat suddenly dragged to a great depth by a whale, was found when drawn up so saturated with water forced into its pores, that it sank in water like a stone for a year afterwards: even sea-water is reduced in bulk from 20 to 19 solid inches at the depth of 20 fathoms. The compression that a whale can endure is wonderful. All fish are capable of sustaining great pressures as well as sudden changes of pressure. Divers in the pearl-fisheries exert great muscular strength; but man cannot bear the increased pressure at great depths, because his lungs are full of air, nor can he endure the diminution of it at great altitudes above the earth.

The depth to which the sun's light penetrates the ocean depends upon the transparency of the water, and cannot be less than twice the depth to which a person can see from the surface. In parts of the Arctic Ocean shells are distinctly seen at the depth of 80 fathoms; and among the West India islands, in 30 fathoms water, the bed of the sea is as clear as if seen in air: shells, corals, and sea-weeds of every hue display the tints of the rainbow.

The purest spring is not more limpid than the water of the ocean: it absorbs all the prismatic colours except that of ultramarine, which, being reflected in every direction, imparts a hue approaching the azure of the sky. The colour of the sea varies with every gleam of sunshine or passing cloud, although its true tint is always the same when seen sheltered from atmospheric influence. The reflection of a boat on the shady side is often of the clearest blue, while the surface of the water exposed to the sun is bright as burnished gold. The waters of the ocean also derive their colour from insects of the infusorial kind, vegetable substances, and minute particles of matter. It is white in the Gulf of Guinea, black round the Maldives; at California the Vermilion Sea is so called on account of the red colours of the infusoria it contains; the same red colour was observed by Magellan at the mouth of the River Plata. The Persian Gulf is called the Green Sea by eastern geographers, and there is a tract of green water off the Arabian coast so distinct that a ship has been seen in green and blue water at the same time. Rapid transitions take place in the Arctic Sea from ultramarine to olive-green, from purity to opacity. These appearances are not delusive, but constant as to place and colour: the green is produced by myriads of minute insects, which devour one another, and are a prey to the whale. The colour of clear shallow water depends upon that of its bed; over chalk or white sand it is apple-green, over yellow sand dark green, brown or black over dark ground, and grey over mud.

The sea is supposed to have acquired its saline principle when the globe was in the act of subsiding from a gaseous state. The density of sea-water depends upon the quantity of saline matter it contains: the proportion is generally about three or four per cent., though it varies in different places; the ocean contains more salt in the southern than in the northern hemisphere, the Atlantic more than the Pacific. The greatest proportion of salt in the Pacific is in the parallels of 22° N. lat. and 17° S. lat.: near the equator it is less; and in the Polar Seas it is least, from the melting of the ice. The saltness varies with the seasons in these regions, and the fresh water, being lighter, is uppermost. Rain makes the surface of the sea fresher than the interior parts, and the influx of rivers renders the ocean less salt at their estuaries: the Atlantic is brackish 300 miles from the mouth of the Amazons. Deep seas are more saline than those that are shallow, and inland seas communicating with the main are less salt, from the rivers that flow into them: to this however the Mediterranean is an exception, occasioned by the great evaporation and the influx of salt currents from the Black Sea and the Atlantic. The water in the Straits of Gibraltar, at the depth of 670 fathoms, is four times as salt as that at the surface.

Fresh water freezes at the temperature of 32° of Fahrenheit; the point of congelation of salt water is lower. As the specific gravity of the water of the Greenland Sea is about 1·02664, it does not freeze till its temperature is reduced to 28½° of Fahrenheit; so that the saline principle preserves the sea in a liquid state to a much higher latitude than if it had been fresh, while it is better suited for navigation by its greater buoyancy. The healthfulness of the sea is ascribed to the mixing of the water by tides and currents, which prevents the accumulation of putrescent matter.

Raised by the moon and modified by the sun in the equatorial seas, the central area of the two oceans is occupied by a great tidal wave, which oscillates continually, keeping time with the returns of the moon, having its motion kept up by her attraction acting at each return. The height of these oceanic tides depends upon the relative position of the sun and moon, and upon their declination and distances from the earth. From the skirts of this oscillating central area, partial tides diverge in all directions, whose velocity depends upon the depth and local circumstances of the sea: these derivative tides are so much influenced by the form of the shore along which they travel that they become of great magnitude in the higher latitudes, while near the centre of the oscillating area the oceanic tide is often very small. The spring-tides rise 50 or 60 feet on some parts of the British coast; in the Bay of Fundy, in Nova Scotia, they rise 60 feet; at St.

Helena they never exceed three feet; and there is scarcely any tide among many of the tropical islands in the Pacific.

At the equator the tide follows the moon at the rate of 1000 miles an hour; but the derivative tides are so retarded by the form of coasts and irregularities at the bottom of the sea, that a tide is sometimes impeded by an obstacle till a second tide reaches the same point by a different course, and the water rises to double the height it would otherwise have attained: a complete extinction of the tide takes place when a high-water interferes in the same manner with a low-water, as in the centre of the German Ocean; and when two unequal tides of contrary phases of rise and fall meet, the greater overpowers the lesser, and the resulting height is equal to their difference: such varieties occur chiefly among islands, and at the estuaries of rivers. When the tide flows suddenly up a river, it checks the descent of the stream, so that a high wave, called a bore, is driven with force up the channel. This sometimes occurs in the Ganges; and in the Amazons, at the equinoxes, during three successive days, five of these destructive waves, from 12 to 15 feet high, follow one another up the river daily. In the Turury Channel, in Cayenne, the sea rises 40 feet in five minutes, and as suddenly ebbs.

There may be some small flow of the water westward with the oceanic tide under the equator, though it is imperceptible; but that does not necessarily follow, since the tide in the open ocean is merely an

alternate rise and fall of the surface, so that the *motion*, not the water, follows the moon. A bird resting on the sea is not carried forward as the waves rise and fall: indeed, if so heavy a body as water were to move at the rate of 1000 miles in an hour, it would cause universal destruction, since in the most violent hurricanes the velocity of the wind hardly exceeds 100 miles an-hour. Over shallows however, and near the land, the water does advance, and rolls in waves on the beach.

The friction of the wind combines with the tides in agitating the surface of the ocean, and, according to the theory of undulations, each produces its effect independently of the other; wind, however, not only raises waves, but causes a transfer of superficial water also. Attraction between the particles of air and water, as well as the pressure of the atmosphere, brings its lower stratum into adhesive contact with the surface of the sea. If the motion of the wind be parallel to the surface, there will still be friction, but the water will be smooth as a mirror; but if it be inclined, in however small a degree, a ripple will appear. The friction raises a minute wave, whose elevation protects the water beyond it from the wind, which consequently impinges on the surface at a small angle: thus, each impulse combining with the other produces an undulation which continually advances.

Those beautiful silvery streaks on the surface of a tranquil sea called catspaws by sailors are owing to a partial deviation of the wind from a horizontal direction. The resistance of the water increases with the strength and inclination of the wind. The agitation at first extends little below the surface, but, in long-continued gales, even the deep water is troubled: the billows rise higher and higher; and as the surface of the sea is driven before the wind, their "monstrous heads," impelled beyond the perpendicular, fall in wreaths of foam. Sometimes several waves overtake one another, and form a sublime and awful sea. The highest waves known are those which occur during a north-west gale off the Cape of Good Hope, aptly called the Cape of Storms by ancient Portuguese navigators; and Cape Horn seems to be the abode of the tempest. The sublimity of the scene, united to the threatened danger, naturally leads to an over estimate of the magnitude of the waves, which appear to rise mountains high, as they are proverbially said to do. There is, however, reason to doubt if the highest waves off the Cape of Good Hope exceed 40 feet from the hollow trough to the summit. They are said to rise 20 feet off Australia, and 16 feet in the Mediterranean. The waves are short and abrupt in small, shallow seas, and on that account are more dangerous than the long rolling billows of the wide ocean.

The undulation called a *ground-swell*, occasioned by the continuance of a heavy gale, is totally different from the tossing of the billows, which are confined to the area vexed by the wind, whereas the ground-swell is rapidly transmitted through the ocean to regions far beyond the direct influence of the gale that raised it; and it continues to heave the smooth and glassy surface of the deep long after the wind and the billows are at rest. A swell frequently comes from a quarter in direct opposition to the wind, and sometimes from various points of the compass at the same time, producing a vast commotion even in a dead calm, without ruffling the surface. They are the heralds that point out to the mariner the distant region where the tempest has howled, and they are not unfrequently the harbingers of its approach. In addition to the other dangers from polar ice, there is always a swell at its margin.

Heavy swells are propagated through the ocean, till they gradually subside from the friction of the water, or till the undulation is checked by the resistance of land, when they roll in surf to the shore, or dash in spray and foam over rocks. The rollers at the Cape de Verde Islands are seen at a great distance approaching like mountains. When a gale is added to a ground-swell, the commotion is great, and the force of the surge tremendous, tossing huge masses of rock and shaking the cliffs to their foundation. The violence of the tempest is sometimes so intense as to quell the billows and blow the water out of the sea, driving it in a heavy shower called *spoon-drift* by sailors. On such occasions saline particles have impregnated the air to the distance of 50 miles inland.

The effect of a gale descends to a comparatively small distance below the surface; the sea is probably tranquil at the depth of 200 or 300 feet: were it not so, the water would be turbid and shell-fish would be destroyed. Anything that diminishes the friction of the wind smooths the surface of the sea: for example, oil, or a small stream of packed ice, which suppresses even a swell. When the air is moist its attraction for water is diminished, and, consequently, so is the friction; hence the sea is not so rough in rainy as in dry weather.

Currents of various extent, magnitude, and velocity disturb the tranquillity of the ocean; some of them depend upon circumstances permanent as the globe itself, others on ever-varying causes. Constant currents are produced by the combined action of the rotation of the earth, the heat of the sun, and the trade winds; periodical currents are occasioned by tides, monsoons, and other periodical winds; temporary currents arise from the tides, melting ice, and from every gale of some duration. A perpetual circulation is kept up in the waters of the main by these vast marine streams. They are sometimes superficial, sometimes submarine, according as their density is greater or less than that of the surrounding sea.

The exchange of water between the poles and the equator gives rise to the great permanent currents in the ocean. Although these depend upon the same causes as the trade winds, they differ essentially in this respect—that, whereas the atmosphere is heated from below by its contact with the earth, and transmits the heat to the strata above, the sea is heated at its surface by the direct rays of the sun, which diminish the specific gravity of the upper strata, especially between the tropics, and also occasion strong and rapid evaporation, both of which causes disturb the equilibrium of the ocean. The rotation of the earth also gives the water a tendency to take an oblique direction in its flow towards the equatorial regions, as, in order to restore the equilibrium, deranged by so many circumstances, great streams perpetually descend from either pole towards the equator. When these currents leave the poles they flow directly north and south; but, before proceeding far, their motion is deflected by the diurnal rotation of the earth. At the poles they have no rotatory motion; and although they gain it more and more by the friction of the water in their progress to the equator, which revolves at the rate of 1000 miles an-hour, they arrive at the tropics before they have acquired the same velocity of rotation with the intertropical ocean. On that account they are left behind, and consequently seem to flow in a direction contrary to the diurnal rotation of the earth. For that reason the whole surface of the ocean, for 30 degrees on each side of the equator, has an apparent tendency from east to west, which produces all the effects of a great current or stream flowing in that direction. The trade winds, which blow constantly nearly the same way, combine to give this current a velocity of 9 or 10 miles in 24 hours.

It is evident that the primary currents, as well as those derived from them, must be subject to periodical variations of intensity of six months' duration, because of the melting of the ice at each pole alternately.

The westerly tendency of the equatorial current in the Atlantic is checked by the continent of America, which stretches directly across its course; so that about the 10th parallel of south latitude it is divided by the coast of Brazil into two branches, one of which runs south and the other north-west. The latter rushes along the coast of Brazil with such force and depth that it is neither deflected by the powerful stream of the river Amazon nor that of the Orinoco. Though much weakened in passing among the West Indian islands, it acquires new strength and the high temperature of 86° of Fahrenheit in the Caribbean Sea. From thence, after sweeping round the Gulf of Mexico, it flows through the States of Florida and along the North American coast to Newfoundland: it is there deflected eastward by the diminished velocity of rotation, and also by a current from Baffin's Bay, so that it proceeds to the Azores. From thence it bends southward, and rejoins the equatorial current, having formed a circuit of 3800 miles with various velocity and a breadth of from 50 to 250 miles, leaving a vast loop or space of water nearly stagnant in its centre, which is thickly covered with sea-weed. The bodies of men, animals, and plants of unknown appearance, brought to the Azores by this stream, suggested to Columbus the idea of land beyond the Western Ocean, and thus led to his discovery of America. The Gulf Stream is more salt, warmer, and of a deeper blue than the rest of the ocean, till it reaches Newfoundland, where it becomes turbid from the shallowness of that part of the sea. Its greatest velocity is 78 miles a-day soon after leaving the Florida Strait, and its greatest breath is 120 miles, though the warm water spreads over the surface of the ocean, to a much greater extent. An important branch leaves this current near Newfoundland, setting towards Britain and Norway, which is again subdivided into many branches, whose origin is recognised by their greater warmth, even at the edge of perpetual ice in the Polar Ocean, while they tend in some degree, by their superficial direction, to prevent the ice from spreading over the North Sea; and in consequence of some of these branches the Spitzbergen Sea is 6° or 7° warmer at the depth of 200 fathoms than it is at the surface. The other branch of the equatorial stream, after setting southward along the coast of Brazil, becomes insensible before reaching the Straits of Magellan.

In the Pacific Ocean a current comes from the south pole along the shores of Chili and Peru to Mexico, having in some seasons a temperature 24° below that of the Equatorial Sea. From Mexico, aided by the equatorial current of the Great Ocean, it crosses the Pacific with so strong a stream, that ships passing from Acapulco to Manilla rarely have occasion to set their sails. Branches flow on each side of Australia, which unite and run through the Bay of Bengal to the extremity of the Indian peninsula; one part then strikes across the ocean, another and greater flows through the Mozambique Channel: these currents then unite in a stream 100 miles broad, and the greater part, called the Lagullus Current, doubles the Cape of Good Hope, and rushes down the coast of Africa till it joins the equatorial current of the Atlantic. These oceanic streams exceed all the rivers in the world in breadth and depth, as well as length. The equatorial current in the Atlantic is 160 miles broad off the coast of Africa, but towards its mid-course, across the Atlantic, its width becomes nearly equal to the whole length of Great Britain; but as it then sends off a branch to the N.W., it is diminished to 200 miles before reaching the coast of Brazil. The depth of this great stream is unknown, but the Brazilian branch must be very profound, since it is not deflected by the river La Plata, which crosses it with so strong a current that its fresh muddy waters are perceptible 500 miles from its mouth. When currents pass over banks and shoals, the colder water rises to the surface, and gives warning of the danger.

The action of these oceanic rivers has been very great on the eastern sides of both continents, where they have scooped out bays and gulfs, and torn off many islands from the land: indeed, the whole earth bears the marks of a great current rushing with violence from the east.

Under-currents are supposed to flow in many places in a direction opposite to the set of the water on the surface, but of these little is known. In summer, the great north polar current coming along the coasts of Greenland and Labrador, together with the current from Davis's Straits, bring icebergs to the margin of the Gulf Stream and disappear. Probably from their density they become under-currents which pass to lower latitudes. Counter-currents on the surface are of such frequent occurrence that there is scarcely a strait joining two seas that does not furnish an example—a current running in along one shore, and a counter-current running out along the other.

Periodical currents are frequent in the eastern seas: one flows into the Red Sea from October to May, and out of it from May to October; in the Persian Gulf this order is reversed. In the Indian Ocean and China Sea the waters are driven alternately backwards and forwards by the monsoons. It is the south-westerly monsoon that causes inundations in the Ganges and a tremendous surf on the coast of Coromandel. The tides also produce periodical currents on the coasts and in straits, the water running in one direction during the flood, and the contrary way in the ebb. The Roost of Sumbury, at the southern promontory of Zetland, runs at the rate of 15 miles an-hour; indeed, the strongest tidal currents known are among the Orkney and Zetland islands; their great velocity arises from local circumstances. Currents in the wide ocean move at the rate of from one to three miles an-hour, and the velocity is less at the margin and bottom of the stream from friction.

Whirlpools are produced by opposing winds and tides: the whirlpool of Maelstrom, on the coast of

Norway, is occasioned by the meeting of tidal currents round the islands of Logodon and Maskoe; it is a mile and a half in diameter, and so violent that its roar is heard at the distance of several leagues.

Although, with winds, tides, and currents, it might seem that the ocean is ever in motion, yet in the equatorial regions, far from land, dead calms prevail; the sea is of the most perfect stillness day after day, rarely does a shower fall, thunder is almost never heard, and the winds are at rest. The sea partakes of the universal quiet, and heaves its low flat waves in noiseless and regular periods, as if nature were asleep.

Salt water is a bad conductor of heat, therefore the temperature of the ocean is less liable to sudden changes than the atmosphere: the influence of the seasons is imperceptible at the depth of 300 feet; and as light probably does not penetrate lower than 700 feet, the heat of the sun cannot affect the bottom of a deep sea. It has been established beyond a doubt by Kotzebue and Sir James Ross, that throughout the whole of the deep ocean the water has an invariable temperature of about 39° 5′ of Fahrenheit at a certain depth depending on the latitude. At the equator the stratum of invariable temperature is at the depth of 7200 feet; from thence it gradually rises till it comes to the surface in S. lat. 56° 26′, where the water has the temperature of 39° 5′ at all depths; it then gradually descends to S. lat. 70°, where it is 4500 feet below the surface.

In going north from the equator the same law is observed: hence with regard to temperature there are three regions in the ocean, one equatorial and two polar. In the equatorial region the temperature of the water at the surface of the ocean is 80°, therefore higher than that of invariable temperature, while in the polar regions it is lower. Thus the surface of the stratum of constant temperature is a curve which begins at the depth of 4500 feet in the southern basin, from whence it gradually rises to the surface in S. lat. 56° 26′; it then sweeps down to 7200 feet at the equator, and rises up again to the surface in the corresponding northern latitude, from whence it descends again to a depth of 4500 feet in the northern basin. From these circumstances Sir James Ross justly infers that the internal heat of the earth has no influence upon the mean temperature of the ocean. The temperature of the surface of the ocean decreases from the equator to the poles. For ten degrees on each side of the line the maximum is 80° of Fahrenheit, and remarkably stable; from thence the decrease to each tropic does not exceed 37°. The tropical temperature would be greater were it not for the currents, because the surface reflects much fewer of the sun's rays, that fall on it directly, than that in higher latitudes, where they fall obliquely. In the torrid zone the surface of the sea is about 35° of Fahrenheit warmer than the air above it, because the polar winds, and the great evaporation which absorbs the heat, prevent equilibrium; and as a great mass of water is slow in following the changes in the atmosphere, the vicissitude of day and night has little influence, whereas in the temperate zones it is perceptible.

The superficial temperature diminishes from the tropics as the latitude increases, more rapidly in the southern than in the northern hemisphere, till towards each pole the sea becomes a solid mass of ice. In the Arctic Ocean the surface is at the freezing point even in summer, and during the eight winter months a continuous body of ice extends in every direction from the pole, filling the area of a circle of between 3000 and 4000 miles in diameter. The outline of this circle, though subject to partial variations, is found to be nearly similar at the same season of each succeeding year, yet there are periodical changes in the polar ice, which are renewed after a series of years. The freezing process itself is a bar to the unlimited increase of the oceanic ice. Fresh water congeals at the temperature of 32° of Fahrenheit, but sea-water must be reduced to 28° 5′ before it deposits its salt and begins to freeze: the salt thus set free, and the heat given out, retard the process of congelation more and more below.

The ice from the north pole comes so far south in winter as to render the coast of Newfoundland inaccessible: it envelops Greenland, sometimes even Iceland, and always invests Spitzbergen and Nova Zembla. As the sun comes north the ice breaks up into enormous masses of what is called packed ice. It is remarkable that in a fine summer the ice suddenly clears away, and leaves an open channel of sea along the western coast of Spitzbergen from 60 to 150 miles wide, extending to 80° or even 80½° N. lat., probably owing to warm currents from low latitudes. In the year 1806 Captain Scoresby forced his ship through 250 miles of packed ice, in imminent danger, until he reached the parallel of 81° 50′, his nearest approach to the pole: the Frozen Ocean is rarely navigable so far.

In the year 1827 Sir Edward Parry arrived at the latitude of 82° 45′, which he accomplished by dragging a boat over fields of solid ice, but he was obliged to abandon the bold and hazardous attempt to reach the pole, because the current drifted the ice southward more rapidly than he could travel over it to the north.

Floating fields of ice 20 or 30 miles in diameter are frequent in the Arctic Ocean; sometimes they extend 100 miles, so closely packed together that no opening is left between them; their thickness, which varies from 10 to 40 feet, is not seen, as there is at least two-thirds of the mass below water. Sometimes these fields, many thousand millions of tons in weight, acquire a rotatory motion of great velocity, dashing against one another with a tremendous collision. Packed ice always has a tendency to drift southwards, even in the calmest weather; and in their progress the ice-fields are rent in pieces by the swell of the sea. It is computed that 20,000 square miles of drift ice are annually brought by the current along the coast of Greenland to Cape Farewell. In stormy weather the fields and streams of ice are covered with haze and spray from constant tremendous concussions; yet our seamen, undismayed by the appalling danger, boldly steer their ships amidst this hideous and discordant tumult.

Huge icebergs are rolled from the glaciers which extend miles from the arctic lands into the sea, especially in Baffin's Bay, and are drifted southwards 2000 miles from their origin to melt in the Atlantic, where they cool the water sensibly for 40 or 50 miles around, and the air to a much greater distance. They vary from a few yards to miles in circumference, and rise hundreds of feet above the surface. Seven hundred such masses have been seen at once in the polar basin. When there is a swell the loose ice dashing against them raises the spray to their very summits; and if a large mass falls from them, they occasionally lose their equilibrium and roll over, causing a swell which breaks up the neighbouring field-ice: the commotion then spreads far and wide, and the uproar resounds for miles like thunder.

Icebergs have the appearance of chalk-cliffs with a glittering surface and emerald-green fractures; pools of water of azure-blue lie on their surface, or fall in cascades into the sea. The field-ice also, and the masses that are heaped up on its surface, are extremely beautiful from the vividness and contrast of their colouring. A peculiar blackness in the atmosphere indicates their position in a fog, and their place and character are shown at night by the reflection of the snow-light on the horizon. An experienced seaman can readily distinguish whether the ice is newly formed, heavy, compact, or open. The blink or snow-light of field-ice is the most lucid, and is tinged yellow; of packed ice it is pure white: ice newly formed has a greyish blink; and a deep yellow tint indicates snow on land.

Icebergs come to a lower latitude by 10° from the south pole than from the north, and appear to be larger. One observed by Captain d'Urville was 13 miles long, with perpendicular sides 100 feet high. They are less varied than those on the northern seas; a tabular form is prevalent. The discovery ships under the command of Sir James Ross met with multitudes bounded by perpendicular cliffs on every side, with flat surfaces from 100 to 180 feet high, sometimes several miles in circumference. On one occasion they fell in with a chain of stupendous bergs close to one another, extending farther than the eye could reach even from the mast-head. Packed ice, too, is in immense quantities: these ships forced their way through a pack 1000 miles broad, often under the most appalling circumstances. It generally consists of smaller pieces than the packs in the comparatively tranquil North Polar seas, where they are often several miles in diameter, and where fields of ice extend beyond the extent of vision. The Antarctic Ocean, on the contrary, is almost always agitated; there is a perpetual swell, and terrific storms are common, which break up the ice and render navigation perilous. The pieces are rarely a quarter of a mile in circumference, and generally much smaller.

A more dreadful situation can hardly be imagined than that of ships beset during a tempest in a dense pack of ice in a dark night, thick fog and drifting snow, with the spray beating perpetually over the decks, and freezing instantaneously. Sir James Ross's own words can alone give an idea of the terrors of one of the many gales which the two ships under his command encountered: "Soon after midnight our ships were involved in an ocean of rolling fragments of ice, hard as floating rocks of granite, which were dashed against them by the waves with so much violence, that their masts quivered as if they would fall at every successive blow; and the destruction of the ships seemed inevitable from the tremendous shocks they received. In the early part of the storm the rudder of the 'Erebus' was so much damaged as to be no longer of any use; and about the same time I was informed by signal that the 'Terror's' was completely destroyed, and nearly torn away from the stern-post. Hour passed away after hour without the least mitigation of the awful circumstances in which we were placed. The loud crashing noise of the straining and working of the timbers and decks, as they were driven against some of the heavier pieces of ice, which all the exertions of our people could not prevent, was sufficient to fill the stoutest heart, that was not supported by trust in Him who controls all events, with dismay; and I should commit an act of injustice to my companions if I did not express my admiration of their conduct on this trying occasion. Throughout a period of

28 hours, during any one of which there appeared to be very little hope that we should live to see another, the coolness, steady obedience, and untiring exertions of each individual, were every way worthy of British seamen.

"The storm gained its height at 2 P.M., when the barometer stood at 28·40 inches, and after that time began to rise. Although we had been forced many miles deeper into the pack, we could not perceive that the swell had at all subsided, our ships still rolling and groaning amidst the heavy fragments of crushing bergs, over which the ocean rolled its mountainous waves, throwing huge masses one upon another, and then again burying them deep beneath its foaming waters, dashing and grinding them together with fearful violence."

For three successive years were these dangers encountered during this bold and hazardous enterprise.

The ocean is one mass of water, which, entering into the interior of the continents, has formed seas and gulfs of great magnitude, which afford easy and rapid means of communication, while they temper the climates of the widely expanding continents.

The inland seas communicating with the Atlantic are larger, and penetrate more deeply into the continents, than those connected with the Great Ocean; a circumstance that gives a coast of 48,000 miles to the former, while that of the Great Ocean is only 44,000. Most of these internal seas have extensive river domains, so that by inland navigation the Atlantic virtually enters into the deepest recesses of the land, brings remote regions into contact, and improves the condition of the less cultivated races of mankind by commercial intercourse with those that are more civilised.

The Baltic, which occupies 125,000 square miles in the centre of northern Europe, is one of the most important of the inland seas connected with the Atlantic; and although inferior to the others in size, the drainage of more than a fifth of Europe flows into it. Only about a fourth part of the boundary of its enormous basin of 900,000 square miles is mountainous; and so many navigable rivers flow into it from the watershed of the great European plain, that its waters are one-fifth less salt than those of the Atlantic: it receives at least 250 streams. Its depth nowhere exceeds 115 fathoms, and generally it is not more than 40 or 50. From that cause, together with its freshness and northern latitude, the Baltic is frozen five months in the year. From the flatness of the greater part of the adjacent country, the climate of the Baltic is subject to influences coming from regions far beyond the limits of its river-basin. The winds from the Atlantic bring warmth and moisture, which, condensed by the cold blasts from the Arctic plains, falls in rain in summer, and deep snow in winter, which also makes the sea more fresh. The tides are imperceptible; but the waters of the Baltic occasionally rise more than three feet above their usual level from some unknown cause—possibly from oscillations in its bed, or from changes of atmospheric pressure.

The Black Sea, which penetrates most deeply into the continent of all the seas in question, has, together with the Sea of Azow, an area of 190,000 square miles; but it must at a remote period have been united with the Caspian Lake, and must have covered all the steppe of Astracan. It receives some of the largest European rivers, and drains about 950,000 square miles; consequently its waters are brackish, and freeze on its northern shores in winter.

Of all the branches of the Atlantic that enter deeply into the bosom of the land, the Mediterranean is the most beautiful and the largest, covering with its dark blue waters more than 760,000 square miles. Situate in a comparatively low latitude, exposed to the heat of the African deserts on the south, and sheltered on the north by the Alps, the evaporation is excessive. Its temperature is 10° or 12° higher than that of the Atlantic. Although its own river domain is only 250,000 square miles, the constant current that sets into it through the Dardanelles brings a great part of the drainage of the Black Sea, so that it is really fed by the melted snow and rivers from the Caucasus, Asia Minor, Abyssinia, the Atlas, and the Alps. Yet the quantity of water that flows into the Mediterranean from the Atlantic by the central current in the Straits of Gibraltar exceeds that which goes out by the lateral ones. In consequence of the excessive evaporation, the water of the Mediterranean is four times as salt as that of the ocean.

The Mediterranean is divided into two basins by

a shallow that runs from Cape Bon on the African coast to the Strait of Messina, on each side of which the water is exceedingly deep, and said to be unfathomable in some parts. This sea is not absolutely without tides: in the Gulf of Venice they rise to three feet, and at the Great Syrte to five at new and full moon; but in most other places they are scarcely perceptible. The surface is traversed by various currents; two of which, opposing one another, occasion the celebrated whirlpool of Charybdis, whose terrors were much diminished by the earthquake of 1783. Its bed is subject to violent volcanic paroxysms; and its surface is studded with islands of all sizes, from the magnificent kingdom of Sicily to mere barren rocks; some actively volcanic, others of volcanic formation, and many of the secondary geological period.

Various parts of its coasts are in a state of great instability; in some places they have sunk down and risen again more than once within historical record.

Far to the north the Atlantic penetrates the American continent by Davis's Straits, and spreads out into Baffin's Bay, twice the size of the Baltic, very deep, and subject to all the rigours of an arctic winter—the very storehouse of icebergs, the abode of the walrus and whale. Hudson's Bay, though without the Arctic circle, is but little less dreary.

Very different is the character of those vast seas where the Atlantic comes "cranking in" between the northern and southern continents of America.

The surface of the sea in Baffin's Bay is seldom above the freezing point; here, on the contrary, it is always 89° of Fahrenheit; while the Atlantic Ocean, in the same latitude, is not above 77° or 78°. Of that huge mass of water partially separated from the Atlantic by a long line of islands and banks, the Caribbean Sea is the largest. It is as long from east to west as the distance between Great Britain and Newfoundland, and occupies a million of square miles. Its depth is very great in many places, and its water limpid. The Gulf of Mexico, fed by the Mississippi, one of the greatest of rivers, is more than half its size, or about 625,000 square miles, so that the whole forms a sea of great magnitude. Its shores, and the shores of the numerous islands, are dangerous from shoals and coral reefs; but the interior of these seas is not. The trade winds prevail there; they are subject to severe northern gales; and some parts are occasionally visited by tremendous hurricanes.

The Pacific does not penetrate the land in the same manner that the Atlantic does the continent of Europe. The Red Sea and Persian Gulf are joined to it by very narrow straits; but almost all the internal seas on the eastern coast of Asia, except the Yellow Sea, are great gulfs shut in by islands, like the Caribbean Sea and the Gulf of Mexico: to which the China Sea (the Toung-Haï), the Sea of Japan, and that of Okhotsk, are perfectly analogous.

The set of the great oceanic currents has scooped out and indented the southern and eastern coasts of the Asiatic continent into enormous bays and gulfs, and has separated large portions of the land, which now remain as islands—a process which probably has been increased by the submarine fires extending along the eastern coast from the equator nearly to the Arctic circle.

The perpetual motion of the ocean by winds, tides, and currents, is continually but slowly changing the form and position of the land—steadily producing those vicissitudes on the surface of the earth to which it has been subject for ages, and to which it will assuredly be liable in all time to come.

CHAPTER XVI.

SPRINGS — BASINS OF THE OCEAN—ORIGIN, COURSE, AND FLOODS OF RIVERS—HYDRAULIC SYSTEMS OF EUROPE—AFRICAN RIVERS: THE NILE, NIGER, ETC.

THE vapour which rises invisibly from the land and water ascends in the atmosphere till it is condensed by the cold into clouds, which restore it again to the earth in the form of rain, hail, and snow: hence there is probably not a drop of water on the globe that has not been borne on the wings of the wind. Part of this moisture restored to the earth is re-absorbed by the air, part supplies the wants of animal and vegetable life, a portion is carried off by streams, and the remaining part percolates through porous soils till it arrives at a stratum impervious to water, where it accumulates in subterranean lakes often of great extent. The mountains receive the greatest portion of the aërial moisture, and, from the many alternations of permeable and impermeable strata they contain, a complete system of reservoirs is formed in them, which, continually overflowing, form perennial springs at different elevations, that unite and run down their sides in incipient rivers. A great portion of the water at these high levels penetrates the earth till it comes to an impermeable stratum below the plains, where it collects in a sheet, and is forced by hydrostatic pressure to rise in springs through cracks in the ground to the surface. In this manner the water which falls on hills and mountains is carried through highly inclined strata to great depths, and even below the bed of the ocean, in many parts of which there are springs of fresh water. In boring artesian wells the water often rushes up with such impetuosity by the hydrostatic pressure as to form jets 40 or 50 feet high. In this operation several successive reservoirs have been met with: at St. Ouen, in France, five sheets of water were found: the water in the four first not being good, the operation was continued to a greater depth. It consists merely in boring a hole of small diameter, and lining it with a tube. It rarely happens that water may not be procured in this way; and as the substratum in many parts of deserts is an argillaceous marl, it is probable that artesian wells might be bored with success.

A spring will be intermittent when it issues from an opening in the side of a reservoir fed from above if the supply be not equal to the waste, for the water will sink below the opening, and the spring will stop till the reservoir is replenished. Few springs give the same quantity of water at all times; they also vary much in the quantity of foreign matter they contain. Mountain springs are generally very pure; the carbonic acid gas almost always found in them goes into the atmosphere, and their earthy matter is deposited as they run along, so that river-water from such sources is soft, while wells and springs in the plains are hard and more or less mineral.

The water of springs takes its temperature from that of the strata through which it passes. Mountain springs are cold, but, if the water has penetrated deep into the earth, it acquires a temperature depending on that circumstance.

The temperature of the surface of the earth varies with the seasons to a certain depth, where it becomes permanent and equal to the mean annual temperature of the air above. It is evident that the depth at which this stratum of invariable temperature lies must vary with the latitude. At the equator the effect of the seasons is imperceptible at the depth of a foot below the surface; between the parallels of 40° and 52° the temperature of the ground in Europe is constant at the depth of from 55 to 60 feet; and in the high Arctic regions the soil is perpetually frozen a foot below the surface. Now, in every part of the world where experiments have been made, the temperature of the earth increases with the depth below the constant stratum at the rate of 1° of Fahrenheit for every 50 or 60 feet of perpendicular depth: hence, should the increase continue to follow the same ratio, even granite must be in fusion at little more than five miles below the surface. In Siberia the stratum of frozen earth is some hundred feet thick, but below that the increase of heat with the depth is three times as rapid as in Europe. The temperature of springs must therefore depend on the depth to which the water has penetrated before it has been forced to the surface either by the hydraulic pressure of water at higher levels or by steam. If it never goes below the stratum of invariable temperature, the heat of the spring will vary with the seasons more or less according to the depth below the surface: should the water come from the constant stratum itself, its temperature will be invariable; and if from below it, the heat will be in proportion to the depth to which it has penetrated. Thus there may be hot and even boiling springs hundreds of miles distant from volcanic action and volcanic strata, of which there are many examples, though they are more frequent in volcanic countries and those subject to earthquakes. The temperature of hot springs is very constant, and that of boiling springs has remained unchanged for ages: shocks of earthquakes sometimes affect their temperature, and have even stopped them altogether Jets of steam of high tension are frequent in volcanic countries, as in Iceland.

Both hot and cold water dissolves and combines with many of the mineral substances it meets with in the earth, and comes to the surface from great depths as medicinal springs, containing various ingredients. So numerous are they that in the Austrian dominions alone there are 1500, and few countries of any extent are destitute of them. They contain sulphuric and carbonic acids, sulphur, iron, magnesia, and other matters. Boiling springs deposit silex, as in Iceland, Italy, and in the Azores; and others of lower temperature deposit carbonate and sulphate of lime in enormous quantities all over the world. Springs of pure brine are very rare; those in Cheshire are rich in salt, and have flowed unchanged 1000 years, a proof of the tranquil state of that part of the globe. Many substances that lie beyond our reach are brought to the surface by springs, as naphtha, petroleum, and borax; petroleum is particularly abundant in Persia, and numberless springs and lakes of it surround some parts of the Caspian Sea. It is found in immense quantities in various parts of the world.

RIVERS.

Rivers have had a greater influence in the location and fortunes of the human race than almost any other physical cause; and since their velocity has been overcome by steam navigation, they have become the highway of the nations.

They frequently rise in lakes which they unite with the sea; in other instances they spring from small elevations in the plains, from perennial sources in the mountains, alpine lakes, melted snow, and glaciers, but the everlasting storehouses of the mightiest floods are the iceclad mountains of table-lands.

Rivers are constantly increased, in descending the mountains and traversing the plains, by tributaries, till at last they flow into the ocean, their ultimate destination and remote origin. "All rivers run into the sea, yet the sea is not full," because it gives in evaporation an equivalent for what it receives.

The Atlantic, the Arctic, and the Pacific Oceans, are directly or indirectly the recipients of all the rivers, therefore their basins are bounded by the principal watersheds of the continents; for the basin of a sea or ocean does not mean only the bed actually occupied by the water, but comprehends also all the land drained by the rivers which fall into it, and is bounded by an imaginary line passing through all their sources. These lines generally run through the elevated parts of a country that divide the streams which flow in one direction from those that flow in another. But the watershed does not coincide in all cases with mountain-crests of great elevation, as the mere convexity of a plain is often sufficient to throw the streams into different directions.

None of the European rivers flowing directly into the Atlantic exceed the 4th or 5th magnitude, except the Rhine; the rest of the principal streams come to it indirectly through the Baltic, the Black Sea, and the Mediterranean. It nevertheless drains nearly half of the old continent, and almost all the new, because the Andes and Rocky Mountains, which form the watershed of the American continent, lie along its western side, and the rivers which rise on the western slope of the Alleghanies are tributaries to the Mississippi, which comes indirectly into the Atlantic by the Gulf of Mexico.

The Arctic Ocean drains the high northern latitudes of America, and receives those magnificent Siberian rivers that originate in the Altaï range from the steppe of the Kerghis to the extremity of Kamtschatka, as well as the very inferior streams of north European Russia. The running waters of the rest of the world merge in the Pacific. The Caspian and Lake of Aral are mere lakes, which receive rivers but emit none.

Mountain-torrents gradually lose velocity in their descent to the low lands by friction, and when they enter the plains their course becomes still more gentle, their beds smoother, and their depth greater. A slope of one foot in 200 prevents a river from being navigable, and a greater inclination forms a rapid or a cataract. The speed, however, does not depend upon the slope alone, but also upon the height of the source of the river, and the pressure of the body of water in the upper part of its course; consequently, under the same circumstances, large rivers run faster than small, but in each individual stream the velocity is perpetually varying with the form of the banks, the winding of the course, and the changes in the width of the channel. The Rhone, one of the most rapid European rivers, has a declivity of one foot in 2620, and flows at the rate of 120 feet in a minute; the sluggish rivers in Flanders have only half that velocity. The Danube, the Tigris, and Indus are among the most rapid of the large rivers.

When one river falls into another, the depth and velocity are increased, but not always proportionally to the width of the channel, which sometimes even becomes less, as at the junction of the Ohio with the Mississippi. When the angle of junction is very obtuse, and the velocity of the tributary stream great, it sometimes forces the water of its primary to recede a short distance. The Arve, swollen by a freshet, occasionally drives the water of the Rhone back into the Lake of Geneva; and it once happened that the force was so great as to make the mill-wheels revolve in a contrary direction.

Instances have occurred of rivers suddenly stopping in their course for some hours, and leaving their channels dry. On the 26th of November, 1838, the water failed so completely in the Clyde, Nith, and Tiviot, that the mills were stopped eight hours in the lower part of their streams. The cause was the coincidence of a gale of wind and a strong frost, which congealed the water near their sources. Exactly the contrary happens in the Siberian rivers, which flow from south to north over so many hundreds of miles; the upper parts are thawed, while the lower are still frozen, and the water, not finding an outlet, inundates the country.

The alluvial soil carried down by streams is gradually deposited as their velocity diminishes; and if they are subject to inundations, and the coast flat, it forms deltas at their mouths. There they generally divide into branches, which often join again, or are united by transverse channels, so that a labyrinth of streams and islands is formed. Deltas are sometimes found in the interior of the continents, at the junction of rivers, exactly similar to those on the ocean, though less extensive.

Tides flow up rivers to a great distance, and to a height far above the level of the sea. The tide is perceptible in the river of the Amazons 576 miles from its mouth, and it ascends 255 miles in the Orinoco.

In the temperate zones rivers are subject to floods from autumnal rains and the melting of the snow, especially on mountain ranges. The Po, for example, spreads desolation far and wide over the plains of Lombardy; but these torrents are as variable in their recurrence and extent as the climate which produces them. The inundations of the rivers in the torrid zone, on the contrary, occur with that regularity peculiar to a region in which meteoric phenomena are uniform in all their changes. These floods are due to the periodical rains which, in tropical countries, follow the cessation of the trade-winds after the equinox of spring and at the turn of the monsoons, and are thus dependent on the declination of the sun, the immediate cause of all these variations. The melting of the snow, no doubt, adds greatly to the floods of the tropical rivers which rise in the high mountain-chains, but it is only an accessory circumstance; for although the snow-water from the Himalaya swells the streams considerably before the rains begin, yet the principal effect is owing to the latter, as the southern face of the Himalaya is not beyond the influence of the monsoon, and the consequent periodical rains, which besides prevail all over the plains of India traversed by the great rivers and their tributaries.

Under like circumstances, the floods of rivers, whose sources have the same tropical latitude, take place at the same season; but the periods of the inundations of rivers on one side of the equator are exactly the contrary of what they are in rivers on the other side of it, on account of the declination of the sun. The flood in the Orinoco is at its greatest height in the month of August, while that of the river of the Amazons, south of the equinoctial line, is at its greatest elevation in March.* The commencement and end of the annual inundations in each river depend upon the mean time of the beginning, and on the duration of the rains in the latitudes traversed by its affluents. The periods of the floods of such rivers as run towards the equator are different from those flowing in an opposite direction; and as the swell requires time to travel, it happens at regular but different periods in various parts of the same river, if very long. The height to which the water rises in the annual floods depends upon the nature of the country, but it is wonderfully constant in each individual river where the course is long; for the inequalities in the quantity of rain in a district drained by any of its affluents is imperceptible in the general flood, and thus the quantity of water carried down is a measure of the mean humidity of the whole country comprised in its basin from year to year. By the admirable arrangement of these periodical inundations, the fresh soil of the moun-

* Baron Humboldt's Personal Narrative.

tains, borne down by the water, enriches countries far remote from their source. The mountains of the Moon, and of Abyssinia, have fertilized the banks of the Nile through a distance of 2500 miles for thousands of years.

When rivers rise in mountains, water communication between them in the upper parts of their course is impossible; but when they descend to the plains, or rise in the low lands, the boundaries between the countries drained by them become low, and the different systems may be united by canals. It sometimes happens, in extensive and very level plains, that the tributaries of the principal streams either unite or are connected by a natural canal, by which a communication is formed between the two basins—a circumstance advantageous to the navigation and commerce of both, especially where the junction takes place far inland, as in the Orinoco and Amazons, in the interior of South America. The Rio Negro, one of the largest affluents of the latter, is united to the Upper Orinoco, in the plains of Esmeralda, by the Cassiquiare—a stream as large as the Rhine, with a velocity of 12 feet in a second. Baron Humboldt observes that the Orinoco sending a branch to the Amazons is, with regard to distance, as if the Rhine should send one to the Seine or Loire. At some future period this junction will be of great importance. These bifurcations are frequent in the deltas of rivers, but very rare in the interior of continents. The Mahomuddy and Gadavery, in Hindostan, seem to have something of the kind, and there are several instances in the great rivers of the Indo-Chinese peninsula.

The hydraulic system of Europe is eminently favourable to inland navigation, small as the rivers are in comparison with those in other parts of the world; but the flatness of the great plain, and the lowness of its watershed, are very favourable to the construction of canals. In the west, however, the Alps and German mountains divide the waters that flow to the Atlantic on one side, and to the Mediterranean and the Black Sea on the other; but in the eastern parts of Europe the division of the waters is merely a more elevated ridge of the plain itself, for in all plains such undulations exist, though often imperceptible to the eye. This watershed begins on the northern declivity of the Carpathian Mountains, about the 23rd meridian, on a low range of hills running between the sources of the Dnieper and the tributaries of the Vistula, from whence it winds in a tortuous course along the plain to the Valday table-land, which is its highest point, 1200 feet above the sea. It then declines northward towards Onega, about the 60th parallel, and lastly turns in a very serpentine line to the sources of the Kama, in the Ural Mountains, near the 62nd degree of north latitude. The waters north of this line run into the Baltic and White Sea, and on the south of it into the Black Sea and the Caspian.

Thus Europe is divided into two principal hydraulic systems; but since the basin of a river comprehends all the plains and valleys drained by it and its tributaries, from its source to the sea, each country is subdivided into as many natural divisions or basins as it has primary rivers, and these generally comprise all the rich and habitable parts of the earth, and are the principal centres of civilization, or are capable of becoming so.

The streams to the north of the general watershed are very numerous; those to the south are of greater magnitude. The systems of the Volga and Danube are the most extensive in Europe: the former has a basin comprising 640,000 square miles, and is navigable throughout the greater part of its course of 1900 miles.

The Danube drains 300,000 square miles, and has 60 navigable tributaries. It rises in the Black Forest at an elevation of 3000 feet above the level of the sea, so that it has considerable velocity, which, as well as rocks and rapids, impede its navigation in many places; but it is navigable downwards, through Austria, for 600 miles to New Orsova, from whence it flows in a gentle current to the Black Sea. The commercial importance of these two rivers is much increased by their flowing into inland seas. By canals between the Volga and the rivers north of the watershed, the Baltic and White Seas are connected with the Black Sea and the Caspian, and the Baltic and Black Sea are also connected by a canal between the Don and the Dnieper. Altogether the water system of Russia is the most extensive in Europe.

The whole of Holland is a collection of deltoid

islands, formed by the Rhine, the Meuse, and the Scheldt; a structure very favourable to commerce, and has facilitated an extensive internal navigation. The Mediterranean is already connected with the North Sea by the junction canal of the Rhone and the Rhine, and this noble system, extended over the whole of France by 7591 miles of canals, has conduced mainly to the improved state of that great country.

Many navigable streams rise in the Spanish mountains: of these the Tagus has depth enough for the largest ships. In point of magnitude, however, many are of the inferior orders, but canals have rendered them beneficial to the country. Italy is less fortunate in her rivers, which only admit of vessels of small burthen. Those in the north are by much the most important, especially the Po and its tributaries, which, by steam-boats, connect Venice and Milan with various fertile provinces of central Italy; but whatever advantages nature has afforded to the Italian states have been improved by able engineers, both in ancient and modern times.

The application of the science of hydraulics to rivers took its rise in northern Italy, which has been carried to such perfection in some points that China is the only country which can vie with it in the practice of irrigation. The lock on canals was in use in Lombardy as early as the 13th century, and in the end of the 14th it was applied to two canals which unite the Ticino to the Adda by that great artist and philosopher Leonardo da Vinci: about the same time he introduced the use of the lock into France.

Various circumstances combine to make the British rivers more useful than many others of greater magnitude. The larger streams are not encumbered with rocks or rapids; they all run into branches of the Atlantic; the tides flow up their channels to a considerable distance; and above all, though short in their course, they end in wide gulfs, capable of containing whole navies—a circumstance that gives an importance to streams otherwise utterly insignificant when compared either with the great rivers of the old or new continent.

The Thames, whose basin is only 5027 square miles, and whose length is but 240 miles, of which however 204 are navigable, spreads its influence over the remotest parts of the earth; its depth is sufficient to admit large vessels even up to London, and throughout its navigable course a continued forest of masts display the flags of every nation; its banks, which are in a state of perfect cultivation, are the seat of the highest civilization, moral and political. Local circumstances have undoubtedly been favourable to this superior development, but the earnest and energetic temperament of the Saxon race has rendered the advantages of their position available. The same may be said of other rivers in the British islands, vying in commercial activity with the Thames. There are 2789 miles of canal in Britain, and, including rivers, 5430 miles of inland navigation, which, in comparison with the size of the country, is very great; it is even said that no part of England is more than 15 miles distant from water communication.

On the whole, Europe is fortunate with regard to its water systems, and its inhabitants are for the most part alive to the bounties which Providence has bestowed.

AFRICAN RIVERS.

In Africa the tropical climate and the extremes of aridity and moisture give a totally different character to its rivers. The most southerly part is comparatively destitute of them, and those that do exist are of inferior size, except the Orange River or Gareep, which has a long course on the table-land, but is nowhere navigable. There is a region of numerous rivers between the 18th degree of south latitude and the equator. They rise in two great watersheds on the table-land, from one of which they go to the Mosambique Channel and Indian Ocean, and from the other they flow to the Atlantic. The first is the range of the mountains of Nyassi, and the high lands that surround the south end of the great lake of that name, 350 miles from Mozambique. From thence all those streams come that flow over the rich plains of Mozambique and Zanguebar. Of these the Zambeze is probably the greatest, and is said to have a course of 900 miles, navigable for 200 or 300 from its mouth. Many other rivers are navigable along this coast, where grain ripens all the year, yielding from 80 to 150 fold, and every eastern production might be raised. The other watershed is a ridge of no great height, that runs from S.E. to N.W. on the table-land west of the dominions of the Zambeze. In it the numerous rivers originate which, after falling in cascades and rapids through the chains that border the table-land on the west, fertilize the luxuriant maritime plains of Benguela, Congo, Angola, and Loango. The Zaire, or Congo, by much the largest of these, is navigable for 140 miles, where the ascent of the tide is stopped by cataracts. The lower course of this river is five or six miles broad, full of islands, and 160 fathoms deep at its mouth. Its upper course, like that of most of these rivers, is unknown; the greater number are fordable on the table-land, but, from the abrupt descent of the high country to the maritime plains, none of them afford access to the interior of south Africa.

The mountainous edge of the table-land, with its terminal projections, Senegambia and Abyssinia, which separate the northern from the southern deserts, is the principal source of running water in Africa. Various rivers have their origin in these mountainous regions, of which the Nile and the Niger yield in size only to some of the great Asiatic and American rivers; in importance and historical interest the Nile is inferior to none.

Two large rivers unite their streams to form the Nile—the Bahr-el-Abiad or White Nile and the

Bahr-el-Azrek or Blue Nile, but the sources and course of the White Nile are yet unknown: it is said to rise in Donga, in the Mountains of the Moon; and the Blue Nile in Abyssinia, in the table-land of Dembea, in the mountains that separate Tigré from Amhara, 10,000 feet above the sea. These two rivers converge during a long and often turbulent course, and unite at last in the plains of Sennaar.

The Tecazze, the largest affluent, issues from the mountains of Lasta, and is the chief river in the kingdom of Tigré. Its affluents fall in cascades from 100 to 150 feet high, and it takes its name of Tecazze or "The Terrible" from the impetuosity with which it rushes through the chasms and over the precipices of the mountains. It joins the main stream in 17° 35′ N. lat., from whence down to the Mediterranean, a distance of 1200 miles, the Nile does not receive a single brook. The first part of the Nile's course is interrupted by cataracts, from the geological structure of the Nubian desert, which consists of a succession of broad sterile terraces, separated by ranges of rocks running east and west. Over these the Nile runs in nine or ten cataracts, the last of which is at Syene, where it enters Egypt. Most of them are only rapids, where each successive fall of water is not a foot high. That they were higher at a former period has recently been ascertained by Dr. Lepsius, the very intelligent traveller, sent by the King of Prussia at the head of a mission to explore that country. He found a series of inscriptions on the rocks in Sennaar, marking the height of the Nile at different periods; and it appears from these that in that country the bed of the river had been 30 feet higher than it is now.

Fifteen miles below Cairo, and 90 miles from the sea, the Nile is divided into two branches, of which, one, running in a northerly direction, enters the Mediterranean below Rosetta; the other, cutting Lower Egypt into two nearly equal parts, enters the sea above Damietta; so that the delta between these two places has a sea-coast of 150 miles.

The basin of the Nile, occupying an area of 500,000 square miles, has an uncommon form: it is wide in Ethiopia and Nubia; but for the greater part of a winding course of 2750 miles it is merely a verdant line of the softest beauty, suddenly and strongly contrasted with the dreary waste of the Red desert. Extending from the equatorial far into the temperate zone, its aspect is less varied than might have been expected on account of the parched and showerless country it passes through. Nevertheless, from the great elevation of the origin of the river, the upper part has a perpetual spring, though within a few degrees of the equator. At the foot of the table-land of Abyssinia the country is covered with dense tropical jungles, while the rest of the valley is rich soil, the detritus of the mountains for thousands of years.

As the mean velocity of the Nile, when not in flood, is about two miles and a half an hour, a particle of water would take twenty-two days and a half to descend from the junction of the Tecazze to the sea; hence the retardation of the annual inundations of the Nile in its course is a peculiarity of this river, owing to some unknown cause towards its origin which affects the whole stream. In Abyssinia and Sennaar the river begins to swell in April, yet the flood is not sensible at Cairo till towards the summer solstice; it then continues to rise about a hundred days, and remains at its greatest height till the middle of October, when it begins to subside, and arrives at its lowest point in April and May. The height of the flood in Upper Egypt varies from 30 to 35 feet; at Cairo it is 23, and in the northern part of the Delta only 4 feet.

Anubis, or Sirius, the Dog-star, was worshipped by the Egyptians, from its supposed influence on the rising of the Nile. According to Champolion, their calendar commenced when the heliacal rising of that star coincided with the summer solstice, the time at which the Nile began to swell at Cairo. Now this coincidence made the nearest approach to accuracy 3291 years before the Christian era; and as the rising of the river still takes place precisely at the same time and in the same manner, it follows that the heat and periodical rains in Upper Ethiopia have not varied for 5000 years. In the time of Hipparchus the summer solstice was in the sign of Leo; and probably about that period the flowing of the fountains from the mouths of lions of basalt and granite was adopted, as emblematical of the pouring forth of the floods of the Nile. The emblem is still common in Rome, though its origin is probably forgotten; and the signs of the zodiac have moved backwards more than 30°.

The two greatest African rivers, the Nile and the Niger, are dissimilar in almost every circumstance; the Nile, discharging itself for ages into a sea, the centre of commerce and civilization, has been renowned by the earliest historians, sacred and profane, for the exuberant fertility of its banks, and for the learning and wisdom of their inhabitants, who have left magnificent and imperishable monuments of their power and genius. It was for ages the seat of science, and by the Red Sea it had intercourse with the most highly cultivated nations of the East from time immemorial. The Niger, on the contrary, though its rival in magnitude, and running through a country glowing with all the brilliancy of tropical vegetation, has ever been inhabited by barbarous or semi-barbarous nations; and its course till lately was little known, as its source still is. In early ages, before the pillars of Hercules had been passed, and indeed long afterwards, the Atlantic coast of Africa was an unknown region; and thus the flowing of the Niger into that lonely ocean kept the natives in their original rude state. Such are the effects of local circumstances on the intellectual advancement of man.

The sources of the Niger, Joliba, or Quorra, are supposed to be on the northern side of the Kong Mountains, in the country of Bambarra, more than 1600 feet above the level of the sea. From thence it goes north, and, after passing through Lake Debo, makes a wide circuit in the plains of Soudan to Timbuctoo, through eight or nine degrees of latitude; then bending round, it again approaches the Kong Mountains, at the distance of 1000 miles in a straight line from its source; and having threaded them, it flows across the low lands into the Gulf of Guinea, a course of 2300 miles. In the plains of Soudan it receives many very large affluents from the high land of Senegambia on the west; and the Chadda on the east—a navigable river larger than itself, the outlet of the great Lake Chad, which drains the lofty Komri, or Mountains of the Moon—falls into it a little below Fandah after a course of some hundred miles: thus it affords an uninterrupted water communication from the Atlantic to the heart of Africa. Long before leaving the plains of Soudan the Niger becomes a noble river, with a smooth stream, gliding at the rate of from five to eight miles an-hour, varying in breadth from one to eight miles. Its banks are studded with densely populous towns and villages, groves of palm-trees and cultivated fields.

This great river divides into three branches near the head of a delta which is equal in area to the whole of Ireland, intersected by navigable branches of the principal stream in every direction. The soil is rich mould, and the vegetation so rank that the trees seem to grow out of the water. The Nun, which is the principal or central branch, flows into the sea near Cape Formosa, and is that which the brothers Lander descended. There are, however, six rivers which run into the Bight of Benin, all communicating with the Niger and with one another. The Old Calabar is the most eastern; it rises in the high land of the Calbongos, and is united to the Niger by a natural canal. The Niger throughout its long winding course lies entirely within the tropic of Cancer, and is consequently subject to periodical inundations, which reach their greatest height in August, about 40 or 50 days after the summer solstice. The plains of Soudan are then covered with water and crowded by boats. These fertile regions are inaccessible to Europeans from the pernicious climate, and dangerous from the savage condition of many of the tribes.

The coast of Guinea west from the Niger is watered by many streams of no great magnitude from the Kong Mountains. The table-land of Senegambia is the origin of the Rio Grande, the Gambia, the Senegal, and others of great size, and also many of an inferior order that fertilise the luxuriant maritime plains on the Atlantic. Their navigable course is cut short by a semicircular chain of mountains which forms the western boundary of the high land through which they thread their way in rapids and cataracts. The Gambia rises in Foula Toro, and after a course of about 600 miles enters the Atlantic by many branches connected by natural channels, supposed at one time to be separate rivers. The Senegal, the largest river in this part of Africa, is

850 miles long. It receives many tributaries in the upper part of its course, and in the lower is full of islands. It drains two lakes, has several accessories, and is united to the basin of the Gambia by the river Neriho.

CHAPTER XVII.

ASIATIC RIVERS—EUPHRATES AND TIGRIS—RIVER SYSTEMS SOUTH OF THE HIMALAYA—CHINESE RIVERS—SIBERIAN RIVERS.

THE only river system of importance in western Asia is that of the Euphrates and Tigris. In the basin of these celebrated streams, containing an area of 230,000 square miles, mounds of rubbish on a desolate plain are the only vestiges that remain of the great cities of Nineveh and Babylon. Innumerable ruins and inscriptions, also records of the glory of times less remote, have been discovered by adventurous travellers, and bear testimony to the truth of some of the most interesting pages of history. The Euphrates, and its affluent the Merad-Chaï, supposed to be the stream forded as the Euphrates by the 10,000 Greeks in their retreat, rise in the heart of Armenia, and, after running 1800 miles on the table-land to 38° 41′ of north latitude, they join the northern branch of the Euphrates, which rises in the Gheul Mountains, near Erzeroum. The whole river then descends in rapids through the Taurus chain, north of Rumkala, to the plains of Mesopotamia.

The Tigris comes from Dearbeker, more to the east, and, after receiving auxiliaries from the high lands of Kourdistan, it pierces the Taurus Mountains at Mosul, and descends rapidly in a tortuous course to the same plains, where it is joined by many streams from the Lusistán Mountains, some of which are navigable, and may ultimately be of great commercial importance. The country through which they flow is extremely beautiful, and rich in corn, date-groves, and forest-trees. Near the city of Bagdad, the two rivers, approaching, surround the plain of Mesopotamia, unite at Koona, and run 150 miles in one stream to the Persian Gulf, under the name of Chat-el-Arab. The banks of the Tigris and Euphrates are quite desolate, alternately vast swamps or burnt up, and in many parts covered with brushwood or grass. The remains of numerous canals, joining these great rivers and their affluents, show the former magnitude of this most ancient water system. The floods of this river are very regular in their periods; they begin in March, and attain their greatest height in June.

The Persian Gulf may be navigated by steam all the year, the Euphrates only eight months; it might however afford easy intercourse with eastern Asia, as it did in former times. The distance from Aleppo to Bombay by the Euphrates is 2870 miles, of which 2700, from Bir to Bombay, are by water; in the time of Queen Elizabath this was the common route to India, and a fleet was then kept at Bir expressly for that navigation.

Five systems of rivers of the first magnitude descend from the central table-land of eastern Asia and its mountain barriers, all different in origin, direction, and character, while they convey to the ocean a greater volume of water than all the rivers of the rest of the continent conjointly. Of these, the Indus, the double system of the Ganges and Brahmapootra, and the group of parallel rivers in the Indo-Chinese peninsula, water the plains of southern Asia; the great system of rivers that descend from the eastern terraces of the table-land irrigates the fertile lands of China; and lastly, the Siberian rivers, not inferior to any in magnitude, carry the waters of the Altaï to the Arctic Ocean.

The hard-fought battles and splendid victories recently gained by British valour over a bold and well-disciplined foe have added to the historical interest of the Indus and its tributary streams, now the boundaries of our Asiatic territories.

The sources of the Indus and Sutlej were only ascertained in 1812: the Ladak, the largest branch of the Indus, has its origin in the snowy mountains of Karakorum; and the Shyook, which is the smaller stream, rises in the Kentese or Kangri range, a branch of the Himalaya, which extends along the table-land of Tibet, west of the sacred lake of Manasarora. These two streams join north-west of Ladak, and form the Indus; the Sutlej, its principal tributary, springs from the sacred lake itself. Both are fed by streams of melted snow from the northern side of the Himayala, and both flow westward along the extensive longitudinal valleys of Tibet. The Sutlej breaks through the Himalaya about the 75th meridian, and traverses the whole breadth of the chain in frightful chasms and clefts in the rocks to the plains of the Punjab; the Indus, after continuing its course on the table-land through several degrees of longitude farther, descends by the Hindoo Coosh, west of the valley of Cashmere, to the same plain. Three tributaries, the Jelum or Hydaspes, the Hydraötes, and the Chenab, all superior to the Rhone in size, flow from the southern face of the Himalaya, and with the Sutlej join the Indus before it reaches Mittum; hence the name Punjab, "the plain of the five rivers," now one of our valuable possessions in the East. From Mittum to the ocean, the Indus, like the Nile, does not receive a single accessory, from the same cause—the sterility of the country through which it passes. The Cabul river, which rises near Guzni, but is joined by a larger affluent from the lofty plain of Pamere, flows along the edge of the Persian table-land, through picturesque and dangerous defiles, and forms the limit between eastern and western Asia. It then joins the Indus at the town of Attock, and is the only tributary of any magnitude that comes from the west.

The Indus is not favourable to navigation: for 70 miles after it leaves the mountains the descent in a boat is dangerous, and it is nowhere navigable for steam-vessels of more than 30 inches draught of water; yet, from the fertility of the Punjab, and the near approach of its basin to that of the Ganges at the foot of the mountains, it must ultimately be a valuable acquisition, and the more especially because it commands the principal roads between Persia and India, one through Cabul and Peshawer to Attock, and the other from Herat through Candahar to the same place. The delta of the Indus, formerly celebrated for its civilization, has long been a desert; but from the vitality of the soil, and the change of political circumstances, it may again resume its pristine aspect. It is 60 miles long, and presents a face of 120 miles to the sea at the Gulf of Oman, where the river empties itself by many mouths, of which only three or four are navigable: one only can be entered by vessels of 50 tons, and all are liable to change. The tide ascends them with extraordinary rapidity for 75 miles, and so great is the quantity of mud carried by it and the absorbing violence of the eddies, that a vessel wrecked on the coast was buried in sand and mud in two tides. The annual floods begin with the melting of the snow in the Himalaya in the end of April, come to their height in July, and end in September. The length of this river is 1500 miles, and it drains an area of 400,000 square miles.

The second group of south Indian rivers, and one of the greatest, is the double system of the Ganges and Brahmapootra. These two rivers, though wide apart at their sources, converge to a common delta, and constitute one of the most important groups on the globe.

Mr. Alexander Elliot, of the Body Guard in Bengal, son of Admiral Elliot, with his friends, are the first who have accomplished the arduous expedition to the sources of the Ganges. The river flows at once in a very rapid stream not less than 40 yards across, from a huge cave in a perpendicular wall of ice at the distance of about three marches from the temple of Gungootree, to which the pilgrims resort. Mr. Elliot says, "The view from the glacier was perfectly amazing; beautiful or magnificent is no word for it—it was really quite astonishing. If you can fancy a bird's-eye view of all the mountains in the world in one cluster, and every one of them covered with snow, it would hardly give you an idea of the sight which presented itself."

Many streams from the southern face of the Himalaya unite at Hurdwar to form the great body of the river. It flows from thence in a south-easterly direction through the plains of Bengal, receiving in its course the tribute of 19 or 20 rivers, of which 12 are larger than the Rhine. About 220 miles in a direct line from the Bay of Bengal, into which the Ganges flows, the innumerable channels and branches into which it splits form an intricate maze over a delta twice as large as that of the Nile.

The sources of the Brahmapootra, a river equal in volume to the Ganges, though not in length, are some hundreds of miles distant from those of the latter. They lie to the north of the Birman empire, but whether they spring from the eastern extremity of the Himalaya or from some snow-clad branch of it is unknown. The upper course of the river among the lofty defiles of the mountains is completely zigzag, but soon after passing through the sacred pool of Brahma-Koond it enters the plains of Upper Assam, and receives the name of Brahmapootra—" the offspring of Brahma;" the natives call it the Lahit, Sanscrit for "red river." In Upper Assam, through which it winds 500 miles and forms some very extensive channel islands, it receives six very considerable accessories, of which the origin is unknown, though some are supposed to come from the table-land of Tibet. They are only navigable in the plains, but vessels of considerable burthen ascend the parent stream as high as Sampura. Before it enters the plains of Bengal, below Goyalpara, the Brahmapootra runs with rapidity in great volume, and, after receiving the river of Bhotan and other streams, its branches unite with those of the Ganges about 40 miles from the coast, but the two rivers enter the sea by different mouths, though they sometimes approach within two miles. The length of the Brahmapootra is probably 860 miles, so that it is 500 miles shorter than the Ganges: the volume of water discharged by it during the dry season is about 146,188 cubic feet in a second; the quantity discharged by the Ganges in the same time and under the same circumstances is only 80,000 cubic feet. In the perennial floods the quantity of water poured through the tributaries of the Brahmapootra from their snowy sources is incredible: the plains of Upper Assam are an entire sheet of water from the 15th of June to the 15th of September, and there is no communication but by elevated causeways eight or ten feet high; the two rivers with their branches lay the plain of Bengal under water for hundreds of miles annually. They begin first to swell from the melting of the snow on the mountains; but before their inferior streams overflow from that cause, all the lower parts of Bengal adjacent to the Ganges and Brahmapootra are under water, from the swelling of these rivers by the rains. The increase is arrested before the middle of August by the cessation of the rains in the mountains, though they continue to fall longer in the plains. The delta is traversed in every direction by arms of the rivers. The Hoogly branch, at all times navigable, passes Calcutta and Chandernagor; and the Hauringotta arm is also navigable, as well as the Ganges properly so called. The channels, however, are perpetually changing, from the strength of the current and the prodigious quantity of matter washed from the high lands; the Ganges alone carries to the sea 600,000 cubic feet of mud in a second, the effects of which are perceptible 60 miles from the coast. The elevation of the mountains, and indeed of the land generally, must have been enormous, since it remains still so stupendous after ages of such degradation. The Sunderbands, a congeries of innumerable river islands formed by the endless streams and narrow channels of the rivers, as well as by the indentations of arms of the sea, line the coast of Bengal for 180 miles—a wilderness of jungle and heavy timber. The united streams of the

Ganges and Brahmapootra drain an area of 650,000 square miles, but there is scarcely a spot in Bengal more than 20 miles distant from a river navigable even in the dry season.

These three great rivers of southern India do not differ more widely in their physical circumstances than in the races of men who inhabit their banks, yet from their position they seem formed to unite nations the most varied in their aspect and speech. The tributaries of the Ganges and Indus come so near to each other at the foot of the mountains, that a canal only two miles long would unite them, and thus an inland navigation from the Bay of Bengal to the Gulf of Oman might be established.

An immense volume of water is poured in a series of nearly parallel rivers of great magnitude and strength through the Indo-Chinese peninsula into the ocean opposite the Sunda Archipelago. They rise in those elevated regions at the south-eastern angle of the table-land of Tibet, the lofty but unknown provinces of the Chinese empire, and water the great valleys that extend nearly from north to south with perfect uniformity, between chains of mountains no less uniform, which spread out like a fan as they approach the sea. Scarcely anything is known of the origin or upper parts of these rivers, and with a few exceptions almost as little of the lower.

Their number amounts to six or seven, all large, though three surpass the rest—the Irriwaddy, which waters the Birman empire, and falls into the Bay of

Bengal at the Gulf of Martaban; the Meinam or river of Siam; and the river of Cambodja, which flows through the empire of Annam: the two last go into the China Sea.

The sources of the Irriwaddy are in the same chain of mountains with those of the Brahmapootra, more to the south. Its course is through countries hardly known to Europeans, but it seems to be navigable by boats before coming to the city of Amarapoora, south of which it enters the finest and richest plain of the empire, containing its four capital cities. There it receives two large affluents, one from the Chinese province of Yunnan, which flows into the Irriwaddy at the city of Ava, 446 miles from the sea, the highest point attained by the British force during the Birmese war.

From Ava to its delta the Irriwaddy is a magnificent river, more than four miles broad in some places, but encumbered with channel islands. In this part of its course it receives its largest tributary, and forms in its delta one of the most extensive systems of internal navigation. The Rangoon is the only one of its 14 mouths that is always navigable, and in it the commerce of the empire is concentrated. The internal communication is extended by the junction of the two most navigable deltoid branches with the rivers Salven and Pegu, by natural canals: that joining the former is 200 miles long; the canal uniting the latter is only serviceable at high water.

The Meinam, one of the largest Asiatic rivers, is

less known than the Irriwaddy: it comes from the Chinese province of Yunnan and runs through the kingdom of Siam, which it cuts into several islands by many diverging branches, and enters the Gulf of Siam by three principal arms, the most easterly of which forms the harbour of Bangkok. It is joined to the Meinam Kong or Cambodja by the small river Anan-Myit.

The river of Cambodja has the longest course of any in the peninsula; it is supposed to be the Lang-thsang, which rises in the high land of K'ham, in eastern Asia, not far from the sources of the great Chinese river, the Yang-tsi-kiang. After traversing the elevated plain of Yunnan, where it is navigable, it rushes through the mountain-barriers; and on reaching a wider valley, about 300 miles from its mouth, it is joined to the Meinam by the natural canal of the Anan-Myit. More to the south it is said to split into branches which unite again.

The ancient capital of Annam is situated on the Cambodja, about 150 miles from the sea: a little to the south its extensive delta begins, projects far into the ocean, and is cut in all directions by arms of the river navigable during the floods; three of its mouths are permanently so for large vessels up to the capital. The Sai-gon, more to the east, is much shorter than the river of Cambodja, though said to be 1000 miles long; but Europeans have not ascended higher than the town of Sai-gon. Near its mouth it sends off several branches to the eastern arm of the Cambodja. All rivers of this part of

Asia are subject to periodical inundations, which fertilize the plains at the expense of the mountains.

The parallelism of the mountain-chains constitutes formidable barriers between the upper basins of the Indo-Chinese rivers, and decided lines of separation between the inhabitants of the intervening valleys; but this inconvenience is in some degree compensated by the natural canals of junction and the extensive water communication towards the mouths of the rivers.

"The Sons of the Ocean," a double system of colossal rivers which drain 1,400,000 square miles of the Chinese empire, rise in the two extensive and principal terraces on the eastern slope of the table-land of central Asia. The length of the Hoang-Ho is 2000 miles, that of the Yang-tsi-kang 2900. Though near at their beginning, they are widely separated north and south, as they proceed on their eastern course, by the mountain-chains that border the table-land; but they again approach, and are not more than 100 miles apart when they enter the Whang-Hai or Yellow Sea. They are united in central China by innumerable canals, and form the grandest and most extensive water system in existence.

The Hoang-Ho brings down in one hour 2,000,000 cubic feet of earth, whence, like the Tiber of old, it is called the "Yellow" River.

Strong tides from the Pacific go up these rivers 400 miles, and for the time prevent the descent of the fresh water, which forms large interior seas frequented by thousands of trading-vessels, and they irrigate the productive lands of central China, from time immemorial the most highly cultivated and the most densely peopled region of the globe.

Almost all the Chinese rivers of less note—and they are numerous—feed these giant streams, with the exception of the Ta-si-kiang, and the Pei-ho, or White River, which have their own basins. The former, rising to the east of the town of Yunnan, flows through the plains of Canton eastward to the Gulf of Canton, into which it discharges itself, increased in its course by the Sekiang.

The White River, rising in the mountains near the great wall, becomes navigable a few miles east of Pekin, unites with the Eu-ho, joins the great canal, and, as the tide ascends it for 80 miles, it is crowded with shipping.

Four great rivers, the Amur, the Lena, the Yenessee, and the double system of the Irtish and Oby, not inferior in size to any rivers in Asia, carry off the waters that come from the Altaï chain, and from the mountains and terraces on the northern declivity of the central table-land. Two of these, the Amur and Lena, rise in the Baikalian mountains, the source of more great rivers than any group of its size. The Amur, the sources of which are partly in the Russian dominions, though its course is chiefly in China, is 2000 miles long, including its windings, and has a basin of 853,000 square miles. Almost all its accessories come from that part of the Baikalian group called the Yablonnoi Khrebit by the

Russians, and Khing-Khan-Oola by the Chinese. The river Onon, which is the parent stream, has its origin in the Khentai Khan, a branch of the latter; and though its course is through an uninhabited country, it is celebrated as being the birthplace and the scene of the exploits of Tshingis Khan. After passing through the lake of Dalai-nor, which is 210 miles in circumference, it takes the name of Argun, and forms the boundary between the Chinese and Russians for 400 miles: it is then joined by the Silka, where it assumes the Tunguse name of the Amur, or Great River; the Mandchoos call it the Saghalia, or Black Water. It receives most of the unknown rivers that come from the mountain-slopes of the Great Gobi, and falls into the Pacific opposite to the island of Tarakaï, after having traversed three degrees of latitude and thirty-three of longitude.

The Lena, whose basin occupies 800,000 square miles, springs from mountains 20 miles west from the Lake of Baikal, and runs north-east through more than half its course to the Siberian town of Yakutzk, the coldest town on the face of the earth, receiving in its course the Witim and the Alekma, its two principal affluents; the former from the Baikal Mountains, the latter from Stannovoi Khrebit, the most southerly part of the Aldan range. North of Yakutzk, about the 63rd parallel of latitude, the Lena receives the Aldan, its greatest tributary, which also comes from the Stannovoi Khrebit: it then goes to the Arctic Ocean, between banks of frozen mud, prodigious masses of which are hurled down by the summer floods, and bring to view the bones of those huge animals of extinct species which at some remote period had found their nourishment in these desert plains. The length of the Lena including its windings is 1900 miles.

A difference in the pressure of the air has been observed on the banks of this river, on the shores of the sea of Okhotsk, and at Kamtschatka, which shows that in the distance of five degrees of latitude there is an *apparent* difference in the level of the sea amounting to 159 feet.* A similar phenomenon was observed by Captain Foster near Cape Horn, and by Sir James Ross throughout the South Polar Ocean.

The Yenessei, a much larger river than the Lena, drains about 1,000,000 square miles, and is formed by the union of the Great and Little Kem. The former rises at the junction of the Sayansk range with the Baikalian mountains to the north-west of Lake Kassagol; the latter comes from the Egtag or Little Altaï, in quite an opposite direction; so that these two meet nearly at right angles, and take the name of Yenessei: it then crosses the Seyansk range in cataracts and rapids, entering the plains of Siberia below the town of Krasnagarsk. Many rivers join it in this part of its course, chiefly the Angora from the Lake Baikal; but its greatest tributaries, the Upper and Lower Tungurka, both large rivers from the Baikalian mountains, join it lower down, the first to the south, the latter to the

* M. Erman.

north of the town of Yeniseisk, whence it runs north to the Icy Ocean, there forming a large gulf, its length measured along its bed being 2500 miles.

The Oby rises in the Lake of Toleskoi, "The Lake of Gold," in Great Tartary; all the streams of the Lesser Altaï unite to swell it and its great tributary the Irtish. The rivers which come from the northern declivity of the mountains go to the Oby, those from the western sides to the Irtish, which springs from numerous streams on the south-western declivity of the Little Altaï, and runs westward into Lake Zainzan, 200 miles in circumference. Issuing from thence it takes a westerly course to the plain on the north of Semissalatinsk. In the plain it is joined by the Tobol, which crosses the steppe of the Kirghiz Cossacks from the Ural Mountains, and soon unites with the Oby: the joint stream then proceeds to the Arctic Ocean in 67° N. lat. The Oby is 2000 miles long, and the basin of these two rivers occupies a third part of Siberia.

Before the Oby leaves the mountains, at a distance of 1200 miles from the Arctic Ocean, its surface has an absolute elevation of not more than 400 feet, and the Irtish, at the same distance, is only 72 feet higher; both are consequently sluggish. When the snow melts they cover the country like seas; and as the inclination of the plains, in the middle and lower parts of their course, is not sufficient to carry off the water, those immense lakes and marshes are formed which characterise this portion of Siberia.

The bed of the Oby is very deep; and there are

no soundings at its mouth: hence the largest vessels might ascend at least to its junction with the Irtish. Its many affluents also might admit ships, did not the climate oppose an insurmountable obstacle the greater part of the year. Indeed, all Siberian rivers are frozen annually for many months, and even the ocean along the Arctic coasts is rarely disencumbered from ice; consequently these vast rivers never can be important as navigable streams; but towards the mountains they afford water communication from the steppe of Issim to the Pacific. They abound in fish and waterfowl, for which the Siberian braves the extremest severity of the climate.

Local circumstances have nowhere produced a greater difference in the human race than in the basins of the great rivers north and south of the table-land of eastern Asia. The Indian, favoured by the finest climate, and a soil which produces the luxuries of life, intersected with rivers navigable at all seasons, and affording easy communication with the surrounding nations, attained early a high degree of civilization; while the Siberian and Samoide, doomed to contend with the rigours of the polar blasts in order to maintain mere existence, have never risen beyond the lowest grade of humanity: but custom softens even the rigour of this stern life, so that here also a share of happiness is enjoyed.

END OF VOL. I.

Printed in the United States
By Bookmasters